Inequalities from Complex Analysis

Complete Set ISBN 0-88385-000-1
Vol. 28 ISBN 0-88385-033-8

Printed in the United States of America

Current Printing (last digit):
10 9 8 7 6 5 4 3 2 1

The Carus Mathematical Monographs

Number Twenty-Eight

Inequalities from Complex Analysis

John P. D'Angelo
University of Illinois

Published and Distributed by
THE MATHEMATICAL ASSOCIATION OF AMERICA

THE
CARUS MATHEMATICAL MONOGRAPHS

Published by
THE MATHEMATICAL ASSOCIATION OF AMERICA

The following Monographs have been published:

1. *Calculus of Variations,* by G. A. Bliss (out of print)
2. *Analytic Functions of a Complex Variable,* by D. R. Curtiss (out of print)
3. *Mathematical Statistics,* by H. L. Rietz (out of print)
4. *Projective Geometry,* by J. W. Young (out of print)
5. *A History of Mathematics in America before 1900,* by D. E. Smith and Jekuthiel Ginsburg (out of print)
6. *Fourier Series and Orthogonal Polynomials,* by Dunham Jackson (out of print)
7. *Vectors and Matrices,* by C. C. MacDuffee (out of print)
8. *Rings and Ideals,* by N. H. McCoy (out of print)
9. *The Theory of Algebraic Numbers,* second edition, by Harry Pollard and Harold G. Diamond
10. *The Arithmetic Theory of Quadratic Forms,* by B. W. Jones (out of print)
11. *Irrational Numbers,* by Ivan Niven
12. *Statistical Independence in Probability, Analysis and Number Theory,* by Mark Kac
13. *A Primer of Real Functions,* third edition, by Ralph P. Boas, Jr.
14. *Combinatorial Mathematics,* by Herbert J. Ryser
15. *Noncommutative Rings,* by I. N. Herstein (out of print)
16. *Dedekind Sums,* by Hans Rademacher and Emil Grosswald
17. *The Schwarz Function and its Applications,* by Philip J. Davis
18. *Celestial Mechanics,* by Harry Pollard
19. *Field Theory and its Classical Problems,* by Charles Robert Hadlock
20. *The Generalized Riemann Integral,* by Robert M. McLeod
21. *From Error-Correcting Codes through Sphere Packings to Simple Groups,* by Thomas M. Thompson
22. *Random Walks and Electric Networks,* by Peter G. Doyle and J. Laurie Snell
23. *Complex Analysis: The Geometric Viewpoint,* by Steven G. Krantz
24. *Knot Theory,* by Charles Livingston
25. *Algebra and Tiling: Homomorphisms in the Service of Geometry,* by Sherman Stein and Sándor Szabó
26. *The Sensual (Quadratic) Form,* by John H. Conway assisted by Francis Y. C. Fung
27. *A Panorama of Harmonic Analysis,* by Steven G. Krantz
28. *Inequalities from Complex Analysis,* by John P. D'Angelo

MAA Service Center
P. O. Box 91112
Washington, DC 20090-1112
800-331-1MAA   FAX: 301-206-9789

Contents

Preface

This book discusses inequalities and positivity conditions for various mathematical objects arising in complex analysis. The inequalities range from standard elementary results such as the Cauchy-Schwarz inequality and the triangle inequality to recent results such as characterizing bihomogeneous polynomials in several variables that are positive away from the origin. The inequalities and positivity conditions in this book form the foundation for a small but beautiful part of complex analysis.

I begin by defining the complex numbers in terms of the real numbers. The prerequisites for starting the book are therefore no more than the elementary undergraduate courses in real analysis and algebra; in order to get something meaningful from the text the reader should also know some linear algebra and elementary complex variable theory. Such readers will find herein complete details of the proofs of numerous useful and interesting theorems in complex analysis and elementary Hilbert space theory. Copious examples and discussions of geometric reasoning aid the reader. The development culminates in complete proofs of recent results in the research literature.

This book is accessible to a wide audience. I have written the first five chapters so that an undergraduate mathematics major who has seen elementary real analysis can read them. Complex variable theory is not a strict prerequisite, although a reader who has never seen complex numbers will probably not get past Chapter I. Many physicists and engineers can read this book; positivity conditions for polynomials often arise in applied mathematics. Detailed discussion of some geometric arguments that have broad scope for application will help readers in this audience. In order to enliven the text, I sometimes interrupt the development with a delightful or novel application of the ideas. I have written the complete text at a level I hope to be accessible to second year graduate students in mathematics.

The starting point for this book is the existence and properties of the real number system $\mathbf{R}$. From there, I define the complex number field $\mathbf{C}$ in Chapter I and prove many results about it. For several reasons the treatment differs considerably from a review of the standard treatment of "one complex variable". One reason is that many texts treat basic complex analysis well, and hence there is no need to repeat certain things here. A second reason is that I wish to anticipate arguments that will arise later in the book. For example, I discuss the absolute value function in great detail, both to prepare for our discussion of the norm on Hilbert space, and to get the reader used to the emphasis on inequalities. I discuss certain simple inequalities in one variable to anticipate more difficult inequalities used later, so that these later inequalities do not jar the reader. This evokes a basic principle from music; a chord containing a C sharp may be used to anticipate a C sharp (appearing as an accidental) that arises later in the melody. Surprises should be pleasant.

Chapter II provides an introduction to complex Euclidean space $\mathbf{C}^n$ and Hilbert spaces. Proofs of the Riesz representation lemma and related results about orthogonal projection appear here. The chapter closes with a discussion of how to use generating functions to verify the orthonormality of a given collection of vectors. This enables us to introduce Laguerre and Hermite polynomials.

Chapter III provides an introduction to functions of several complex variables. I discuss enough theory of holomorphic functions of several variables to introduce the Bergman kernel function and to compute it for the unit ball. Chapter III also includes a nice treatment of the Euler Beta function in n variables.

Chapter IV focuses on that part of linear algebra concerned with positive definite Hermitian forms. I prove the spectral theorem for Hermitian linear operators in finite dimensions in Chapter IV. I defer the proof for compact Hermitian operators on a Hilbert space to Chapter V. Chapter IV includes a careful proof that a finite-dimensional operator is positive definite if and only if its leading principal minor determinants are positive numbers. The proof interprets the minor determinants in terms of sums of squares, and thus anticipates some discussion in Chapters VI and VII. The book has occasional *applications*. In Chapter IV, for example, we show that if $\{x_j\}$ is a collection of distinct positive numbers, then the matrix whose j, k-th entry is the reciprocal of $x_j + x_k$ must be positive definite.

Many classical inequalities (Hadamard, Minkowski, etc.) follow from the results in this chapter, and some appear as exercises. Chapter IV closes with two sections on elementary Fourier analysis, and includes several beautiful inequalities. Hilbert's inequality, the Herglotz theorem, and one form of Wirtinger's inequality appear here.

Chapter V includes a detailed discussion of compact operators on Hilbert space. In particular I prove a simple proposition that interprets compactness as an inequality, and I use this to prove some of the standard results about compact operators. The chapter also includes a section on integral operators, and it closes with an introductory glimpse at singular integral operators. For example, I give a concrete definition of a fractional derivative operator, and then indicate how pseudodifferential operators provide a nice approach to this notion.

Chapter VI considers a point of view I like to call nonlinear Hermitian linear algebra. Consider a real-valued polynomial of several complex variables. The coefficients of such a polynomial may be identified with a Hermitian matrix; we want to apply, in this nonlinear setting, the linear algebra developed in the book. One goal is to understand polynomials whose values are nonnegative. What is the relationship of this condition to the nonnegative definiteness of the matrix of coefficients? To study this question I introduce eight positivity conditions and discuss the implications among them. I provide many examples. Also appearing here is a somewhat unusual discussion of plurisubharmonicity, the complex variable analogue of convexity. This includes a surprising result: for nonnegative bihomogeneous functions, plurisubharmonicity is equivalent to logarithmic plurisubharmonicity. I know of no real-variables analogue of this statement.

The mathematics of Chapter VII arises from the following complex variables analogue of Hilbert's seventeenth problem. Suppose that a polynomial in several complex variables takes on positive values on a set. Must it agree with the quotient of squared norms of holomorphic polynomial mappings there? I have studied this question in recent research papers, both jointly with David Catlin and on my own. Theorem VII.1.1, a stabilization result, is one of the primary goals of the book. The proof here (due to Catlin and myself) applies compact and integral operators in a somewhat unexpected way. The theorem was proved earlier by Quillen using a different but intriguingly related method.

Theorem VII.1.1 gives a decisive answer to the question relating positivity of the values of a polynomial to positivity conditions on the matrix of coefficients. It leads to three additional results about writing nonnegative polynomials as quotients of squared norms. I also apply it to study proper holomorphic mappings between balls in different dimensions.

It is possible to reinterpret some of the results of Chapters VI and VII in the language of holomorphic line bundles. I had originally planned on closing the book with such a chapter. This material would introduce the reader to a modern point of view whose value in mathematics and mathematical physics continues to develop. After soliciting opinions from other mathematicians, I decided not to include this chapter. Doing so would open many doors but not adequately close them. I therefore end the book with a brief afterword, where I mention some generalizations and give references to the research literature.

The book has many exercises; they are numbered beginning with 1 in each chapter. All other items are numbered by stating the kind of item, the number of the chapter, the number of the section, and the item number within the section. For example, Theorem II.3.2 is the second item in Section 3 of Chapter II.

I acknowledge the contributions of many people. Most important are those of my wife Annette and our four small children. Their love has inspired me throughout.

From the mathematical point of view I owe a great debt to David Catlin, whose understanding of most of these ideas far exceeds my own. It was his idea to involve the Bergman kernel in the study of positivity conditions for polynomials. By pursuing this idea we were able to do more than I had originally expected.

I want to specifically acknowledge four mathematicians who read lengthy parts of various preliminary versions, provided comments, and went far beyond the call of duty. Dan Grayson and Alex Isaev read much of the material without expecting anything in return. Their superb comments played a major role in my revision. Harold Boas and Ken Ross read preliminary versions as members of the Committee on Carus Monographs, and each spotted a large number of potential improvements. I owe much to all four of these people. Of course I take full responsibility for the remaining flaws.

Many other mathematicians have provided useful comments on either the mathematics herein or on preliminary versions of the manuscript. They include Salah Baouendi, David Berg, Steve Bradlow, Mike Christ, Charlie Epstein, Fritz Haslinger, Xiaojun Huang, Bob Jerrard, Robert Kaufman, J. J. Kohn, Steve Krantz, László Lempert, Jeff McNeal, Anoush Najarian, Bruce Reznick, Linda Rothschild, Alberto Scalari, Ragnar Sigurddson, Emil Straube, and Alex Tumanov. I wish to specifically thank Steve Krantz for suggesting that I consider writing a Carus monograph. I also want to thank Kang-Tae Kim for inviting me to KSCV4 (the 4th Korean Several Complex Variables meeting) in 1999. My lecture notes for that meeting helped me get started on this book.

I wish to acknowledge the NSF for research support, and IAS, MSRI, AIM, and ESI for profitable time spent there. Spending time at these institutes has been invaluable for me; this has developed my taste and aspirations within mathematics. I took a sabbatical from teaching in order to write this book, and partially supported my family with money I received from the 1999 Bergman prize. I therefore wish to acknowledge the AMS for awarding me this prize. I wish to thank the MAA staff and especially Beverly Ruedi for their efforts in turning my computer files into a finished book. Finally, I acknowledge the contribution of my home institution, the University of Illinois, for providing me an excellent library and computer facilities to help me write this book.

John P. D'Angelo
Urbana, Illinois

CHAPTER I

Complex Numbers

Mathematical analysis requires a thorough study of inequalities. This book will discuss inequalities and related positivity conditions arising from complex analysis, in both one and several variables. The statements, proofs, and applications of these inequalities involve diverse parts of mathematics. We aim to provide a systematic development of this material, beginning with the definition of the complex number field.

In Section I.2 we shall note that the mathematical statement $z \leq w$ does not make sense when z and w are complex but not real numbers; ultimately the inequalities we use in complex analysis must be inequalities about *real* numbers. Furthermore, many important inequalities in this book involve mathematical objects such as linear transformations or polynomials. Such inequalities also rest upon the notion of positive real number.

We therefore begin with the real number system **R**, which is a *complete ordered field*. We will review what this means in order to get started.

I.1 The real number system

Our starting point is the existence of a complete ordered field **R** called the real number system. The *field* axioms allow us to perform algebraic operations; the *order* axioms allow us to manipulate inequalities; the *completeness* axiom enables us to have a good theory of limits. See Appendices A.1 and A.2 for a list of the axioms for a complete ordered field. Here we give an informal discussion.

To say that **R** is a field means that **R** is equipped with operations of addition and multiplication satisfying the usual algebraic laws. This assumption includes the existence of the additive identity 0 and the multiplicative identity 1.

To say that **R** is ordered guarantees that, given real numbers x and y, exactly one of the three statements $x = y$, $x < y$, $x > y$ is true; this is known as the *trichotomy* property. Furthermore we have $1 > 0$. Let x be a real number. We call x *positive* when $x > 0$, we call x *negative* when $x < 0$, and we call x *nonnegative* when $x \geq 0$. In an ordered field, it is an axiom that when x and y are positive numbers, the sum $x + y$ and the product xy are positive numbers. It then follows that sums and products of nonnegative numbers are nonnegative. It also follows that the sum of (any collection of) negative numbers is negative, but that the product of two negative numbers is positive. In particular, for all $x \in \mathbf{R}$, we have $x^2 \geq 0$.

We pause to glimpse one of the main ideas in this book. Squares of real numbers are nonnegative; sums of nonnegative numbers are non-negative, and the quotient of positive numbers is positive. Given mathematical objects about which we can make sense of these words, we ask whether a nonnegative object must be a sum of squares, or perhaps a quotient of sums of squares. We consider this question for linear transformations and for polynomials in several complex variables. There are many subtle points. For example, different positivity conditions apply in different situations. Even after determining the appropriate notion of positivity, the collection of "nonnegative" but not "positive" objects might consist of many things. By contrast the only nonnegative real number that is not positive is 0.

Completeness for the real number system guarantees that the properties of limits learned in calculus are valid. The completeness axiom for the real numbers is often given in terms of least upper bounds. It implies that every Cauchy sequence of real numbers has a limit in

R. We recall the definitions of Cauchy sequence (in greater generality) and of *complete metric space* a bit later.

It is a standard fact, and not relevant to our purposes, that any two complete ordered fields are isomorphic. It is possible to start with the natural numbers **N** as given, to construct from **N** a complete ordered field in a prescribed manner, and to call the result **R**. The construction of the real number system does not concern us here. On the other hand, assuming that **R** is known to us, we will give several constructions of the complex numbers **C**.

I.2 Definition of the complex number field

Given the existence of a complete ordered field **R**, our first goal is to define the field **C** of complex numbers. In this section we define **C** by equipping the Cartesian plane $\mathbf{R}^2$ with operations of addition and multiplication. In Section I.4 we sketch two alternative approaches.

Consider $\mathbf{R}^2 = \mathbf{R} \times \mathbf{R}$, the Cartesian product of **R** with itself. Let (x, y) and (a, b) be elements of $\mathbf{R}^2$. We define addition and multiplication on $\mathbf{R}^2$ by

$$(x, y) + (a, b) = (x + a, y + b)$$

$$(x, y)(a, b) = (xa - yb, xb + ya).$$

We define **C** to be $\mathbf{R}^2$ equipped with these operations of addition and multiplication. After some motivation and discussion of notation we will show that **C** is a field.

It is natural (Exercise 3) to identify the real number x with the complex number $(x, 0)$; this identification enables us to say for example that a given complex number is "real." The definition of addition agrees with the usual definition of vector addition in the real plane. The definition of multiplication arises from the desire to find a square root of the real number -1. We define the *imaginary unit i* to be the complex number $(0, 1)$. Then $i^2 = (0, 1)(0, 1) = (-1, 0)$.

Since $(x, y) = (x, 0) + (0, 1)(y, 0) = x + iy$, it is both natural and efficient to denote the complex number (x, y) by $x + iy$. We can then rewrite the definition of multiplication:

$$(x + iy)(a + ib) = (xa - yb) + i(xb + ya). \tag{1}$$

Note that (1) results from formally applying the distributive law to the product on the left and then writing $i^2 = -1$. After verifying the field axioms for **C** we will dispense with the ordered pair notation and consider (1) to be the definition of multiplication.

It is time to justify our claim that **C** is a field. Observe that $(0, 0)$ is an additive identity, and that $(1, 0)$ is a multiplicative identity. Observe that both addition and multiplication are commutative:

$$(x, y) + (a, b) = (a, b) + (x, y)$$

$$(x, y)(a, b) = (a, b)(x, y).$$

Simple computations (Exercise 4) show that both operations are associative:

$$(x, y) + ((a, b) + (s, t)) = ((x, y) + (a, b)) + (s, t)$$

$$(x, y)((a, b)(s, t)) = ((x, y)(a, b))(s, t).$$

The distributive law relating addition and multiplication holds:

$$(x, y)((a, b) + (c, d)) = (x, y)(a, b) + (x, y)(c, d).$$

It is clear that $(-x, -y)$ is the additive inverse of (x, y). It is less clear what is the multiplicative inverse of a non-zero (x, y). Once we have established the existence of multiplicative inverses for all such (x, y), we will have shown that **C** is a field. We postpone the discussion of inverses until we have introduced several useful concepts.

Henceforth we will write $z = x + iy$ for the complex number (x, y). We call x the *real part* of z and y the *imaginary part* of z. We write $x = \text{Re}(z)$ and $y = \text{Im}(z)$. The notation and terminology are consistent with writing $(x, 0)$ as x and $(0, y)$ as iy. We have already noted that the imaginary unit i satisfies $i^2 = -1$. Our discussion so far enables us to add and multiply complex numbers in the usual way:

$$(x + iy) + (a + ib) = (x + a) + i(y + b)$$
$$(x + iy)(a + ib) = xa + i^2 yb + i(xb + ya)$$
$$= xa - yb + i(xb + ya).$$

To complete the justification that $\mathbf{C}$ is a field it remains to show that a nonzero z has a reciprocal. One can write down an unmotivated formula for the multiplicative inverse. Instead we first introduce the notions of complex conjugation and absolute value. We then derive and better understand the formula for the multiplicative inverse.

Definition I.2.1. For $z \in \mathbf{C}$, with $z = x + iy$, we write $\overline{z}$ for the complex number $x - iy$. The number $\overline{z}$ is called the *complex conjugate* of z, and the function $z \to \overline{z}$ is called *complex conjugation*.

Conjugation plays a vital role, partly because it enables us to obtain in a natural way real-valued expressions from complex variables. See Exercise 2 and Remark I.2.3. We can also use conjugation to define the Euclidean norm of a complex number.

Definition I.2.2. The *absolute value* or *Euclidean norm* $|z|$ of a complex number z is defined by the formula

$$|z| = \sqrt{z\overline{z}}.$$

When $z = x + iy$ we have $x^2 + y^2 = z\overline{z}$, so $z\overline{z}$ is a nonnegative number, and hence has a unique nonnegative square root in the real number system.

Notice that $z = 0$ if and only if $|z| = 0$. Later we will interpret $|z|$ as the distance from z to the origin.

The next exercise implies that the mathematical expression "$z \geq w$" makes no sense for complex numbers z and w unless both happen to be real.

Exercise 1. Show that **C** cannot be made into an *ordered* field. Suggestion: If **C** were ordered, then either i or $-i$ would be positive. From either it follows that -1 is positive, which leads to a contradiction.

Exercise 2. Show that conjugation has the following properties:

1) $\overline{z + w} = \overline{z} + \overline{w}$.
2) $\overline{zw} = \overline{z}\,\overline{w}$.
3) $\overline{z} = z$ if and only if z is real. In particular $\overline{0} = 0$ and $\overline{1} = 1$.
4) $|\overline{z}| = |z|$.

Exercise 3. Prove that the collection of complex numbers $(x, 0)$ for x a real number is a field isomorphic to **R**.

Exercise 4. Verify the associative law for multiplication of complex numbers.

Remark I.2.3. Statements 1) and 2) of Exercise 2 show that complex conjugation is an *automorphism* of the complex number field. It maps **C** to itself and preserves the field operations. Statement 4) shows that it is also an *isometry*. A consequence of these statements is that complex analysis would be unchanged if we used $-i$ as the imaginary unit instead of using i.

Computations involving z and its conjugate $\overline{z}$ are generally more aesthetic than those involving the real and imaginary parts of z; Remark I.2.3 provides a nice explanation. We illustrate by offering two proofs of the simple but fundamental identity

$$|zw| = |z|\,|w|. \tag{2}$$

In each case we prove that $|zw|^2 = |z|^2|w|^2$.

Proof of (2) using real variables: Put $z = x + iy$ and $w = a + ib$. Then

$$|zw|^2 = |(xa - yb) + i(xb + ya)|^2$$

$$= (xa - yb)^2 + (xb + ya)^2$$
$$= x^2a^2 + y^2b^2 + x^2b^2 + y^2a^2$$
$$= (x^2 + y^2)(a^2 + b^2) = |z|^2 |w|^2.$$

Proof of (2) using conjugation:

$$|zw|^2 = zw\overline{zw} = zw\overline{z}\overline{w} = z\overline{z}w\overline{w} = |z|^2 |w|^2.$$

The proof based on conjugation is much more appealing! It uses only the definition of absolute value, the second property from Exercise 2, and the commutativity of multiplication. By contrast, the proof based on the real and imaginary parts requires superfluous algebraic manipulation; we must expand a product, simplify the result, and then factor.

These aesthetic considerations explain why we generally work with complex variables and their conjugates rather than with their real and imaginary parts. It is nonetheless fundamental to note that the real and imaginary parts of z have simple formulas in terms of z and $\overline{z}$:

$$x = \mathrm{Re}(x + iy) = \mathrm{Re}(z) = \frac{z + \overline{z}}{2}$$
$$y = \mathrm{Im}(x + iy) = \mathrm{Im}(z) = \frac{z - \overline{z}}{2i}.$$

We are finally prepared to find multiplicative inverses. Suppose $z \neq 0$; then $|z|^2 \neq 0$. Since $|z|^2$ is real and nonzero its reciprocal $\frac{1}{|z|^2}$ as a real number exists; earlier we noted that we may identify a real number with a complex number. Therefore formula (3) below makes sense. The *reciprocal* (or multiplicative inverse) of z is defined by

$$\frac{1}{z} = \overline{z}\frac{1}{z\overline{z}} = \frac{\overline{z}}{|z|^2}. \tag{3}$$

In terms of x and y we have

$$\frac{1}{z} = \frac{x - iy}{x^2 + y^2} = \frac{x}{x^2 + y^2} + i\frac{-y}{x^2 + y^2}.$$

Either formula directly implies that $z \frac{1}{z} = 1$. We have now finished our demonstration that the complex numbers form a field. The reader should notice that, as in the discussion after Remark I.2.3, the formula for a reciprocal using complex conjugation is simpler and more inspiring than the formula using real and imaginary parts.

We pause to make a few simple observations about this approach to complex analysis. We prefer expressing things in terms of z and $\overline{z}$ rather than in terms of x and y. Often we will view z and $\overline{z}$ as independent variables; this might seem strange! If we know z, then we know $\overline{z}$. It is the point in $\mathbf{R}^2$ obtained by reflecting across the x axis. Nevertheless there is a precise sense in which z and $\overline{z}$ are independent, and this lies behind a huge amount of research in complex analysis.

Application I.7.2 provides a simple circumstance when we may regard z and $\overline{z}$ as independent variables. Suppose p is a polynomial in two complex variables and $p(z, \overline{z})$ vanishes for all z. Then $p(z, w)$ vanishes for all z and w. The result remains true for convergent power series.

I.3   Elementary complex geometry

We next express some familiar geometric concepts and subsets of the Euclidean plane $\mathbf{R}^2$ using complex numbers. This section contains no proofs, and it is somewhat informal. For example, when discussing angles we mention the cosine function, which we do not officially *define* until Section I.7. In a careful logical development we could first define the cosine function as in I.7, define π in terms of it, verify that the cosine function is a bijection from the interval $[0, \pi]$ to the interval $[-1, 1]$, prove that the right-hand side of (4) below lies in $[-1, 1]$, and then finally *define* angles using (4). We want however to develop our complex-geometric intuition, and this minor informality helps more than it hinders.

Distance.   The Euclidean distance between complex numbers z and w is $|z - w|$.

This definition agrees with the usual notion of Euclidean distance between the points (x, y) and (s, t), when $z = x + iy$ and $w = s + it$. In particular, if we view a complex number z as a vector in the plane based at the origin and ending at z, then $|z|$ represents the length of this vector. If u and v are vectors, and we view them as complex numbers in this way, then there is a nice formula for their Euclidean inner product:

$$u \cdot v = \operatorname{Re}(u\bar{v}).$$

Angles. Let u and v be nonzero vectors in the plane with the same base point. Recall that the cosine of the angle θ between u and v satisfies

$$\cos(\theta) = \frac{u \cdot v}{|u||v|}. \tag{4}$$

Hence, when u and v are unit vectors, the cosine of the angle between them satisfies

$$\cos(\theta) = \operatorname{Re}(u\bar{v}).$$

In particular, we obtain a criterion for perpendicularity (orthogonality). The vectors u and v are perpendicular if and only if $u\bar{v}$ is purely imaginary. We further note that u and iu are orthogonal; multiplication of a complex number by i rotates it by ninety degrees.

Circles and disks. Let R be a nonnegative real number, and let $p \in \mathbf{C}$. The *circle* of radius R centered at p is the set of all points z satisfying $|z - p| = R$. The unit circle S^1 is the set of complex numbers z with $|z| = 1$. The *closed disk* of radius R centered at p is the set of all z satisfying $|z - p| \le R$, and the *open disk* of radius R centered at p is the set of all z satisfying $|z - p| < R$.

Lines. Recall that parametric equations of a line through (x_0, y_0) with direction vector (a, b) are

$$(x(t), y(t)) = (x_0 + at, y_0 + bt).$$

The parameter t is a real number, and $(a, b) \neq (0, 0)$. Expressing the parametric equations in complex notation, with $z_0 = x_0 + iy_0$ and $w = a + ib$, shows that an equation of the line through z_0 with direction vector (a, b) is

$$z(t) = z_0 + (a + ib)t = z_0 + wt.$$

Application I.3.1. Find the (minimum) distance to the origin from the line given by $z(t) = z_0 + wt$. Assume that $w \neq 0$. Give a geometric interpretation.

To solve this problem we minimize the squared distance $\delta^2(t)$ to the origin, by calculus. Here

$$\delta^2(t) = |z_0 + wt|^2 = |z_0|^2 + 2t\mathrm{Re}(z_0\overline{w}) + t^2|w|^2.$$

This quadratic polynomial in the real variable t is minimized when its derivative vanishes. Thus we set

$$0 = 2\mathrm{Re}(z_0\overline{w}) + 2t|w|^2$$

and solve for t. Putting

$$t = \frac{-\mathrm{Re}(z_0\overline{w})}{|w|^2}$$

and using $\mathrm{Re}(\zeta) = \frac{\zeta + \overline{\zeta}}{2}$ shows that

$$|z_0 + wt|^2 = \left| \frac{z_0}{2} - \frac{\overline{z}_0 w}{2\overline{w}} \right|^2,$$

and hence the minimum distance from the line to the origin is

$$\left| \frac{z_0}{2} - \frac{\overline{z}_0 w}{2\overline{w}} \right|. \tag{5}$$

This beautiful formula shows that the distance to the origin equals the distance between two specific points; neither is the origin nor on the line! These points are $\frac{z_0}{2}$ and $\frac{\overline{z}_0 w}{2\overline{w}}$; the second can be found geometrically as follows. First find $\frac{z_0}{2}$, and locate its complex conjugate by reflection. Then rotate by multiplying by the complex number $\frac{w}{\bar{w}}$ of absolute value 1. One could also write this number of absolute value 1 as $\frac{w^2}{|w|^2}$.

Example I.3.2. Consider the line given by $z(t) = (2 + 4i) + (1 + i)t$. Its distance to the origin is $\sqrt{2}$, realized at the point $-1 + i$. To see this conclusion using (5), we note that $\frac{w^2}{|w|^2} = i$, $\frac{z_0}{2} = 1 + 2i$, and therefore the two points arising in (5) are $1 + 2i$ and $i(1 - 2i) = 2 + i$. The distance between them is also $\sqrt{2}$.

Remark I.3.3. The parallelogram law. For all complex numbers z and w:

$$|z + w|^2 + |z - w|^2 = 2|z|^2 + 2|w|^2. \tag{6}$$

The proof of (6) is immediate after expanding the left-hand side. The meaning and applications of (6) are more interesting than its proof. Formula (6) says that the sum of the squared lengths of the diagonals of a parallelogram equals the sum of the squared lengths of the four sides.

Application I.3.4. Suppose that z and ζ are distinct complex numbers, and w is a complex number. Then the average of the squared distances from z and ζ to w exceeds the squared distance of the midpoint $\frac{z+\zeta}{2}$ to w.

Proof. The parallelogram law applied to $\frac{z-w}{2}$ and $\frac{w-\zeta}{2}$ yields

$$\left|\frac{z-\zeta}{2}\right|^2 + \left|\frac{z+\zeta}{2} - w\right|^2 = 2\left|\frac{z-w}{2}\right|^2 + 2\left|\frac{\zeta-w}{2}\right|^2. \tag{7}$$

Since $\left|\frac{z-\zeta}{2}\right|^2 > 0$, we obtain

$$\left|\frac{z+\zeta}{2} - w\right|^2 < \frac{|z-w|^2 + |\zeta-w|^2}{2},$$

which gives the desired result. $\qquad\square$

An analogue in higher dimensions of the following corollary of the parallelogram law will arise in the study of projections.

Corollary I.3.5. If z and ζ are distinct points in $\mathbf{C}$, and equidistant to w, then their midpoint is closer to w.

Proof. Let $d^2 = |z - w|^2 = |\zeta - w|^2$ denote the equal squared distances, and let d_m^2 denote the squared distance from the midpoint $\frac{z+\zeta}{2}$ to w. Then (7) yields

$$d_m^2 = \left| \frac{z + \zeta}{2} - w \right|^2 = d^2 - \left| \frac{z - \zeta}{2} \right|^2 < d^2,$$

which gives the desired conclusion. $\square$

I.4   Alternative definitions of the complex numbers

Worthwhile mathematical concepts generally admit various interpretations; different points of view allow for different insights and enhance our understanding. The complex numbers provide a good example. There is no best possible *definition* of $\mathbf{C}$.

In this section we consider two alternative ways for defining the complex numbers. The first uses two-by-two matrices of real numbers; it is motivated by the geometric meaning of complex multiplication. The second uses the notion of the quotient of a ring by an ideal; it is motivated by a desire to manipulate complex numbers as we do real numbers while remembering that $i^2 = -1$. This section can be skipped without affecting the formal development of the book.

I.4.1   Using matrices

Multiplying complex numbers has several geometric interpretations. Consider the definition of multiplication in $\mathbf{C}$ given in Section I.2. Fix

$w = a + ib$ and let $z = x + iy$. The map $z \rightarrow zw$ defines a linear transformation from $\mathbf{R}^2$ to itself; it sends (x, y) into $(ax - by, bx + ay)$. This simple observation suggests identifying complex numbers with linear transformations of $\mathbf{R}^2$ of this particular form. We therefore consider the collection of two-by-two matrices

$$\begin{pmatrix} a & -b \\ b & a \end{pmatrix} \tag{8}$$

where a and b are real numbers. It is easy to see that the sum of two such matrices has the same form, and that the product of two such matrices also does. Note furthermore that the rules for adding and multiplying these matrices correspond precisely to our rules for finding sums and products of complex numbers. Thus we may interpret the complex numbers as the collection of such two-by-two matrices, where the field operations are given by matrix addition and multiplication. From this point of view, we are identifying (8) with the complex number $a + ib$. In particular, the imaginary unit i corresponds to the matrix

$$\begin{pmatrix} 0 & -1 \\ 1 & 0 \end{pmatrix},$$

and real numbers correspond to real multiples of the identity matrix. If z corresponds to

$$\begin{pmatrix} x & -y \\ y & x \end{pmatrix},$$

then $\overline{z}$ corresponds to

$$\begin{pmatrix} x & y \\ -y & x \end{pmatrix}.$$

Their matrix product is $x^2 + y^2$ times the identity matrix. Thus $z\overline{z}$ serves as a definition of $|z|^2$ in this approach as well. Finally, the multiplicative inverse corresponds to the matrix inverse because

$$z^{-1} = \begin{pmatrix} x & -y \\ y & x \end{pmatrix}^{-1} = \frac{1}{x^2 + y^2} \begin{pmatrix} x & y \\ -y & x \end{pmatrix} = \frac{1}{|z|^2} \overline{z}.$$

We summarize the discussion. The collection of two-by-two matrices of real numbers of the form (8), with the field operations given by matrix addition and matrix multiplication, is a field isomorphic to **C**. Finally, we remark without elaboration that this point of view is often useful in complex differential geometry.

I.4.2 Using polynomials

A third approach to the definition of the field of complex numbers considers the ring of polynomials $\mathbf{R}[t]$. Elements of this ring are polynomials with real coefficients in the indeterminate t, and are added and multiplied as usual. The subset I of polynomials divisible by $t^2 + 1$ forms an ideal in $\mathbf{R}[t]$, so we may form the quotient ring $\mathbf{R}[t]/(t^2+1)$. Because the polynomial $t^2 + 1$ is irreducible, the ideal I is a *maximal* ideal. By an elementary result in ring theory the quotient ring is therefore a field.

Taking the quotient amounts to setting $t^2 = -1$. From this point of view, a complex number is an equivalence class of polynomials in one variable t. Each equivalence class contains a unique representative $a + bt$ of degree at most one. The equivalence class containing $a + bt$ *is* the complex number we have been writing $a + ib$. In this case one can easily verify directly (without mentioning maximal ideals) that the quotient ring is a field; it is isomorphic to the complex numbers as defined in Section 2. This definition provides a precise framework for setting $i^2 = -1$ when multiplying complex numbers:

$$(x + iy)(a + ib) = xa + i^2 yb + i(xb + ya)$$
$$= xa - yb + i(xb + ya).$$

We summarize the discussion. The quotient ring $\mathbf{R}[t]/(t^2 + 1)$ is a field under the operations of addition and multiplication arising from $\mathbf{R}[t]$, and it is isomorphic to **C** as defined in Section 2.

I.5 Completeness

We have shown that $\mathbf{C}$ is a field. There is no way to make $\mathbf{C}$ into an ordered field; see Exercise 1. On the other hand, familiar properties of limits easily extend to the complex domain. The crucial idea is to interpret $|z|^2$ as the squared distance of z to the origin. With this notion of distance, $\mathbf{C}$ is a *complete* field.

We defined $\mathbf{C}$ by introducing field operations on $\mathbf{R}^2$. Now we wish to introduce the Euclidean topology on $\mathbf{R}^2$ and obtain the complex Euclidean line. The word *line* is appropriate because we will often consider the field of complex numbers to be a one-dimensional vector space over itself.

Recall that a metric space (M, d) is a set M together with a function $d : M \times M \to \mathbf{R}$ (called the *distance* function) satisfying the following axioms. See for example [Ah], [F], or [TBB] for more information on metric spaces.

Definition I.5.1. Axioms for the distance function d on a metric space (M, d).

1) For each $x, y \in M$ we have $d(x, y) \geq 0$, and $d(x, y) = 0$ if and only if $y = x$.

2) For each $x, y \in M$ we have $d(x, y) = d(y, x)$.

3) (The triangle inequality) For each $x, y, z \in M$ we have

$$d(x, y) + d(y, z) \geq d(x, z).$$

The definition of limit for sequences in a metric space is crucial for doing analysis.

Definition I.5.2. Let (M, d) be a metric space. Let $\{z_\nu\}$ be a sequence in M. Then $\{z_\nu\}$ *converges* to z, or "has limit z," if, for each positive real number ϵ, there is a positive integer N such that

$$\nu \geq N \Rightarrow d(z_\nu, z) \leq \epsilon.$$

A sequence in M is a *Cauchy sequence* if its terms are eventually arbitrarily close together. The following basic definition makes this intuitive notion precise.

Definition I.5.3. Let (M, d) be a metric space. Let $\{z_\nu\}$ be a sequence in M. Then $\{z_\nu\}$ is a *Cauchy sequence* if, for each positive real number ϵ, there is a positive integer N such that

$$\nu, \mu \geq N \Rightarrow d(z_\nu, z_\mu) \leq \epsilon.$$

Definition I.5.4. A metric space (M, d) is *complete* if every Cauchy sequence in M has a limit in M.

Taking $M = \mathbf{R}$ and $d(x, y) = |x - y|$ yields the usual definition of a Cauchy sequence of real numbers. We require several fundamental notions about the real numbers. First is the standard result that a bounded monotone sequence converges. Next are the notions of lim inf and lim sup. Suppose that $\{a_n\}$ is a bounded sequence of real numbers. We define two new sequences by $b_k = \inf_{n \geq k}\{a_n\}$ and $c_k = \sup_{n \geq k}\{a_n\}$. Then each of these sequences is bounded and monotone, and hence has a limit. We write $\liminf(a_n) = \lim_k(b_k)$ and $\limsup(a_n) = \lim_k(c_k)$. Assuming that $\{a_\nu\}$ is bounded, these limits exist and

$$\liminf(a_n) \leq \limsup(a_n).$$

Equality holds if and only if $\lim(a_n)$ exists, in which case all three numbers are equal. In general we allow the values $\pm\infty$ for the lim sup and lim inf of an unbounded sequence.

We return to $\mathbf{C}$. The notions of lim inf and lim sup do not make sense in this setting, but the notion of Cauchy sequence is unchanged. The following result is an analogue of the completeness axiom for $\mathbf{R}$.

Proposition I.5.5. The function $(z, w) \rightarrow |z - w| = d(z, w)$ from $\mathbf{C} \times \mathbf{C} \rightarrow \mathbf{R}$ is a distance function that makes $(\mathbf{C}, d)$ into a complete metric space.

Proof. To show that d defines a distance function we need to show the following:

1) $|z - w| \geq 0$ for all z and w, and equality occurs only when $z = w$.
2) $|z - w| = |w - z|$.
3) The triangle inequality holds.

Statements 1) and 2) are evident. To prove 3) we must verify

$$|z - w| \leq |z - \zeta| + |\zeta - w|. \tag{9}$$

Statement (9) follows from the triangle inequality (10) for the Euclidean norm

$$|\eta + \tau| \leq |\eta| + |\tau|, \tag{10}$$

by putting $\eta = z - \zeta$ and $\tau = \zeta - w$. So we prove (10). To do so it suffices to prove that

$$|\eta + \tau|^2 \leq (|\eta| + |\tau|)^2 = |\eta|^2 + 2|\eta||\tau| + |\tau|^2.$$

Since $|\eta + \tau|^2 = |\eta|^2 + \eta\bar{\tau} + \bar{\eta}\tau + |\tau|^2$ and $\eta\bar{\tau} + \bar{\eta}\tau = 2\text{Re}(\eta\bar{\tau})$, it suffices to verify

$$|\eta\bar{\tau}| = |\eta||\tau|$$

and

$$|\text{Re}(z)| \leq |z|.$$

The first follows from (2) and property 4) of Exercise 2; the second is obvious after squaring.

It remains to prove completeness. Suppose that $\{z_\nu\}$ is a Cauchy sequence of complex numbers. The inequality $|z_\nu - z_\mu| < \epsilon$ implies

$$|\text{Re}(z_\nu) - \text{Re}(z_\mu)| < \epsilon$$

and

$$|\operatorname{Im}(z_\nu) - \operatorname{Im}(z_\mu)| < \epsilon.$$

Therefore the real and imaginary parts of a Cauchy sequence of complex numbers define Cauchy sequences of real numbers. Since $\mathbf{R}$ is complete, they have limits x and y. To finish we must verify that $\{z_\nu\}$ converges to $z = x + iy$. This statement follows from the triangle inequality; we have

$$|z_\nu - z| \le |\operatorname{Re}(z_\nu) - \operatorname{Re}(z)| + |\operatorname{Im}(z_\nu) - \operatorname{Im}(z)| < \epsilon$$

as long as both $|\operatorname{Re}(z_\nu) - \operatorname{Re}(z)|$ and $|\operatorname{Im}(z_\nu) - \operatorname{Im}(z)|$ are less than $\frac{\epsilon}{2}$; this happens for ν sufficiently large.   □

Exercise 5. Suppose that z and w are in the closed unit disk.

1) Prove the inequality

$$|z - w| \le |1 - z\overline{w}|$$

and determine when equality holds.

2) Prove the inequality

$$1 - |w|^2 \le 2|1 - z\overline{w}|.$$

Suggestion: Start with $1 = |1 - z\overline{w} + z\overline{w}|$, or find a geometric proof.

I.6   Convergence for power series

The completeness of $\mathbf{C}$ will be used many times in this book. As a crucial example of its use, we discuss convergence for power series in one variable.

Suppose, for each $n \ge 0$, we are given a complex number c_n. We may then consider the expression

$$a(z) = \sum_{n=0}^{\infty} c_n z^n = \lim_{N \to \infty} \sum_{n=0}^{N} c_n z^n.$$

We call such an infinite sum a *power series* in z. The finite sum $\sum_{n=0}^{N} c_n z^n$ is called the N-th *partial sum* of the series $a(z)$. By definition $a(z)$ *converges* at z if the limit of the sequence of partial sums there exists. Consider the set A consisting of those z for which the series $a(z)$ converges. Then $a : A \to \mathbf{C}$ defines a function. In order to study such functions we want to determine the set A.

Perhaps the most important example is the *geometric series* $\sum_{n=0}^{\infty} z^n$. In this case we can explicitly find the partial sum. In case $z = 1$ the N-th partial sum is $N + 1$, which is the number of terms in the sum. When $z \neq 1$ we have the following explicit formula; its elementary verification is left to the reader:

$$\sum_{n=0}^{N} z^n = \frac{1 - z^{N+1}}{1 - z}. \tag{11}$$

From (11) we see that the limit of the partial sums of the geometric series exists if and only if $|z| < 1$; with that restriction we have

$$\sum_{n=0}^{\infty} z^n = \frac{1}{1 - z}. \tag{12}$$

Exercise 6. Give two proofs of (11). Prove it directly by induction, and also prove it by multiplying the sum by $1 - z$ and using the distributive law. Where must you use induction in the second proof? Verify (12), including a proof that the limit of the partial sums of the geometric series exists if and only if $|z| < 1$.

The geometric series arises for many reasons and in diverse parts of mathematics. It even arises in elementary economics! The proof below of the root test indicates how the geometric series provides a basis of comparison for other series. Other uses will be apparent later. The following result (essentially from freshman calculus) determines the largest open set on which a given series converges. The idea of the proof is to compare the terms of the given series to those of a geometric series.

Proposition I.6.1. Let $a(z) = \sum_{n=0}^{\infty} c_n z^n$ be a power series in z. Either $a(z)$ converges for all z, or there is a nonnegative number R such that $a(z)$ converges for all z with $|z| < R$, and diverges for all z with $|z| > R$. The number R is determined as follows:

Let $L = \limsup(|c_n|^{1/n})$. Put $R = \frac{1}{L}$ when $0 < L < \infty$, and put $R = 0$ when $L = \infty$. We write $R = \infty$ when $L = 0$, corresponding to the case where $a(z)$ converges for all z.

Proof. We consider only the case where $0 < L < \infty$, leaving the other cases to the reader. First we fix z satisfying $|z| < R = \frac{1}{L}$, and verify convergence at z. The idea is to find r with $r < 1$ such that $|c_n z^n| \leq r^n$ for all large n. We then compare the given series with the geometric series. To find r, we first define δ by $L|z| = 1 - \delta$ and then define ϵ by $\epsilon = L\delta$. By definition of the lim sup, there is an N such that $n \geq N$ implies $|c_n| \leq (L + \epsilon)^n$. Hence, with $r = 1 - \delta^2$,

$$|c_n z^n| \leq (L + \epsilon)^n \left(\frac{1 - \delta}{L} \right)^n = (1 + \delta)^n (1 - \delta)^n$$

$$= (1 - \delta^2)^n = r^n.$$

For $n \geq N$, the absolute values of the terms of $a(z)$ are thus dominated by the terms r^n of a convergent geometric series. Hence the full series $a(z)$ converges.

On the other hand, when $|z| > R = \frac{1}{L}$, we put $L|z| = 1 + \delta$. Choose a positive ϵ with $(1 - \frac{\epsilon}{L}) \geq \frac{1}{1+\delta}$. By definition of lim sup, there are infinitely many n for which $|c_n| \geq (L - \epsilon)^n$. For such n we have

$$|c_n z^n| \geq (L - \epsilon)^n \left(\frac{1 + \delta}{L} \right)^n$$

$$= \left(\frac{L - \epsilon}{L} \right)^n (1 + \delta)^n \geq (1 + \delta)^{-n}(1 + \delta)^n = 1.$$

Thus the terms in the series do not tend to zero, and therefore the series diverges. $\square$

The number R is the *radius of convergence* of the series. In many examples one can compute R using the familiar *ratio test*. When $\lim_{n\to\infty} \frac{|c_{n+1}|}{|c_n|}$ exists, this limit equals L as defined in Proposition I.6.1.

Example I.6.2. Consider the series

$$\sum_{n=1}^{\infty} \frac{n^n z^n}{n!}.$$

Using the ratio test, and then the definition from calculus that $e = \lim_{n\to\infty}(1 + \frac{1}{n})^n$, we see that $R = \frac{1}{e}$. □

More generally, we may consider series centered (or based) at p rather than at 0:

$$a(z) = \sum_{n=0}^{\infty} c_n(z - p)^n.$$

In this case we see that the series converges at z when $|z - p| < R$ and diverges at z when $|z - p| > R$. We say that the disk $\{z : |z-p| < R\}$ is the *region of convergence* of $a(z)$. The series may converge at some points z for which $|z - p| = R$. Determining the precise set of such points is a difficult matter that will not arise in this book. It is standard but somewhat misleading to call $\{z : |z - p| = R\}$ the circle of convergence of $a(z)$.

Exercise 8 gives an improvement of the ratio test, and suggests other renditions of it.

Exercise 7. Prove Proposition I.6.1 in the cases where $L = 0$ and $L = \infty$.

Exercise 8. Let $\{a_n\}$ be a sequence of positive numbers. Show that $\sum a_n$ converges if and only if there is a sequence $\{b_n\}$ of positive numbers such that $\sum b_n$ converges and

$$\frac{a_{n+1}}{a_n} \leq \frac{b_{n+1}}{b_n}.$$

Use this test to show that $\sum a_n$ converges when $\frac{a_{n+1}}{a_n} = 1 - \frac{p}{n}$ for $p > 1$, or, more generally, when the ratio behaves asymptotically like $1 - \frac{p}{n}$ and $p > 1$. Give an example where the ratio $\frac{a_{n+1}}{a_n}$ behaves asymptotically like $1 - \frac{1}{n}$ and the series diverges.

The basic objects of interest in complex analysis are functions that can be represented as convergent power series. The region of convergence of a power series is an open disk. In a metric space we write $B_r(p)$ for the set of points x satisfying $d(x, p) < r$ and call $B_r(p)$ the *open ball* of radius r about p. A subset Ω of a metric space is called *open* if, for each $p \in \Omega$, there is some $r > 0$ such that $B_r(p) \subset \Omega$. An open disk in $\mathbf{C}$ is an open subset. We next give the fundamental definition of complex analytic function on an arbitrary open subset of $\mathbf{C}$.

Definition I.6.3. Let Ω be an open set in $\mathbf{C}$, and suppose $f : \Omega \to \mathbf{C}$ is a function. We say that f is *holomorphic* (or *complex analytic*) on Ω if the following is true at each p in Ω. There is an open disk $B_r(p)$ in Ω, and complex numbers c_n for $n \geq 0$, such that

$$f(z) = \sum_{n=0}^{\infty} c_n (z - p)^n$$

for all $z \in B_r(p)$.

Thus, for each p, a holomorphic function is given near p by a power series centered at p which converges in some $B_r(p)$. It is a standard fact in complex analysis that we may choose the radius r of $B_r(p)$ to be the distance from p to the boundary of Ω. One way to prove this is first to represent f via Cauchy's integral formula, and then to expand the term $\frac{1}{\zeta - z}$ that arises there in a geometric series. See [Ah] or [GK]. Although we do not need this fact, mentioning it is valuable because it illustrates the sense in which the geometric series models all holomorphic functions.

We define the derivative f' of a convergent power series f by differentiating term by term. One of the main points of elementary complex variable theory is that this definition is equivalent to the definition using difference quotients. See for example [Ah] or [GK]. It is therefore natural to write $\frac{df}{dz}$ for f'. Thus, when $f(z) = \sum c_n(z - p)^n$ we put

$$f'(z) = \sum_{n=0}^{\infty} nc_n(z - p)^{n-1} = \sum_{n=0}^{\infty} (n + 1)c_{n+1}(z - p)^n. \qquad (13)$$

This definition relies on a crucial fact. If a power series converges for $|z - p| < r$, then its derivative also converges for $|z - p| < r$, and hence the derivative of a holomorphic function is also holomorphic. This fact follows from Proposition I.6.1 and the equality

$$\limsup \left(((n + 1)|c_{n+1}|)^{\frac{1}{n}} \right) = \limsup(|c_n|^{\frac{1}{n}}). \qquad (14)$$

We may now repeatedly differentiate a power series inside its circle of convergence. As usual we write $f^{(k)}(z)$ for the k-th derivative of f at z.

Warning I.6.4. The derivative of a finite sum of differentiable functions is the sum of the derivatives of the terms, but the analogous statement fails in general for infinite sums. An infinite sum is a *limit* of finite sums, and the derivative of a limit of functions need not be the limit of the derivatives. For convergent power series, the derivative of an infinite sum truly is the sum of the derivatives of the terms in the sum. Thus (13) is the natural definition, but its validity relies on (14) and Proposition I.6.1.

Exercise 9. Prove statement (14). Also give an example of a convergent sequence of differentiable functions on the real line for which the derivative of the limit is not the limit of the derivatives.

Exercise 10. Let $\{z_n\}$ be a sequence of complex numbers. Define a new sequence $\{\sigma_n\}$ by

$$\sigma_n = \frac{1}{n} \sum_{j=1}^{n} z_j.$$

1) Suppose $\lim_{n \to \infty} z_n = L$. Prove that $\lim_{n \to \infty} \sigma_n = L$.

2) Give an example of a sequence $\{z_n\}$ for which $\lim_{n \to \infty} z_n$ does not exist, but $\lim_{n \to \infty} \sigma_n$ does exist.

The numbers σ_n are called the *Cesàro means* of the sequence $\{z_n\}$. Convergence in the Cesàro sense thus generalizes convergence in the usual sense from Definition I.5.2. Cesàro convergence arises especially when z_n is the n-th partial sum of a series. In case the Cesàro means of the partial sums of a series converge, one says that the series is *Cesàro summable*. Cesàro means play an important role in Fourier series. See Section IV.7.

I.7   Trigonometry

Complex analysis makes possible a deep understanding of trigonometry. By using polar coordinates to represent a point in **C**, one discovers profound connections among many aspects of elementary mathematics.

Let z be a nonzero complex number. Then there is a unique point ω on the unit circle, namely $\omega = \frac{z}{|z|}$, such that ω is a positive multiple of z. Elementary trigonometry suggests we can write $\omega = \cos(\theta) + i \sin(\theta)$. This formula presumes that we have given precise definitions of the cosine and sine functions. Many elementary books are a bit sloppy about these definitions; a definition involving angles requires a precise definition of angle. To reconcile this matter, we proceed differently. We have used completeness to prove needed results about convergent sequences and series. Now (Exercise 11) we can find the radius of convergence of each power series in (15) below using the ratio test. We obtain infinity in each case. Hence, for $z \in$ **C**, we define

$$\exp(z) = e^z = \sum_{n=0}^{\infty} \frac{z^n}{n!} \qquad (15)$$

$$\cos(z) = \sum_{n=0}^{\infty} \frac{(-1)^n}{(2n)!} z^{2n}$$

$$\sin(z) = \sum_{n=0}^{\infty} \frac{(-1)^n}{(2n+1)!} z^{2n+1}.$$

Exercise 11. Prove that these series converge for all z in **C**.

Remark I.7.1. Instead of defining the sine and cosine functions by their Taylor series, it is possible to define them directly in terms of the exponential function:

$$\cos(z) = \frac{e^{iz} + e^{-iz}}{2}$$

$$\sin(z) = \frac{e^{iz} - e^{-iz}}{2i}.$$

We chose the power series definition because these two formulas might seem unmotivated.

When z is real, the power series definitions for the exponential, cosine, and sine functions agree with their Taylor series expansions from elementary calculus. The power series thus extend three important elementary functions to the complex domain. For us, however, the series give the definitions of the functions, and the familiar facts about them are not yet available to us. It is a delightful and easy exercise to derive them. We sketch some of this development.

In Exercise 13 you are asked to verify, when z is complex:

$$e^{iz} = \cos(z) + i\sin(z).$$

It follows from Exercise 12 below that $1 = e^{iz}e^{-iz}$. Combining these two formulas yields

$$1 = e^{iz}e^{-iz} = (\cos(z) + i\sin(z))\,(\cos(z) - i\sin(z))$$
$$= \cos^2(z) + \sin^2(z).$$

Thus we obtain the basic trigonometric identity:

$$\cos^2(z) + \sin^2(z) = 1. \tag{16}$$

The exponential and trigonometric functions satisfy the usual differential equations from calculus. According to the discussion near (13), we may differentiate the power series term by term to discover the formula $\frac{d}{dz}e^z = e^z$. Also $\frac{d}{dz}\sin(z) = \cos(z)$ and $\frac{d}{dz}\cos(z) = -\sin(z)$.

Exercise 12. Verify the crucial identity

$$e^{z+w} = e^z e^w. \tag{17}$$

Suggestion 1: Replace z by $z + w$ in (15) and use the binomial theorem. Pay attention to the indices of summation. Suggestion 2: Put $f(z) = e^{z+w}$. Show that $f' = f$ and therefore that $f(z) = f(0)e^z$. Make sure you don't use what you are trying to prove!

Exercise 13. (de Moivre's formula) Verify, for z complex, that

$$e^{iz} = \cos(z) + i\sin(z). \tag{18}$$

Conclude de Moivre's formula: for y real and n an integer we have

$$\cos(ny) + i\sin(ny) = (\cos(y) + i\sin(y))^n.$$

Notice that the exponential, cosine, and sine functions take real values on the real line. When t is real, formula (18) therefore exhibits the real and imaginary parts of e^{it}. As a consequence of (16), e^{it} lies on the unit circle.

It is possible to prove (16) without using complex numbers. The starting point is the definition of the cosine and sine functions of a real variable via their power series. Direct computation and knowledge that we can rearrange the terms of a power series inside the region of convergence combine to give (16). Calculus provides the following more elegant approach.

From the theory of elementary analysis, the power series for cosine and sine define differentiable functions, and their derivatives are given by termwise differentiation of their power series. The usual relationships for their derivatives follow:

$$\frac{d}{dt}\sin(t) = \cos(t)$$

$$\frac{d}{dt}\cos(t) = -\sin(t).$$

Note that $\sin(0) = 0$ and $\cos(0) = 1$. Let $\phi(t) = (\cos(t))^2 + (\sin(t))^2$. Then $\phi(0) = 1$. By differentiating and using the formulas for the derivatives, we see that the derivative $\phi'(t)$ vanishes for all real t, and hence ϕ is constant. Thus (16) holds.

We return to the complex setting. Applying formulas (17) and (18) yields the beautiful result:

$$e^z = e^{x+iy} = e^x e^{iy} = e^x(\cos(y) + i\sin(y)).$$

Every trigonometric identity is a consequence of the formulas (17) and (18). The complex exponential also enables us to introduce polar coordinates. We call (19) the *standard polar representation* of the nonzero complex number z:

$$z = |z|e^{i\theta}. \qquad (19)$$

In (19), θ is the real number with $0 \le \theta < 2\pi$ such that $\frac{z}{|z|} = \cos(\theta) + i\sin(\theta)$.

Other representations arise by adding an integer multiple of 2π to θ. In solving equations one must generally consider *all* such representations. See Exercise 15.

Exercise 14. Use de Moivre's formula (Exercise 13) and the binomial theorem to find formulas for $\cos(nt)$ and $\sin(nt)$ in terms of $\cos(t)$ and $\sin(t)$.

Exercise 15. Find all complex cube roots of i.

The next exercise is used in the proof of Application I.7.2 and also anticipates some of the ideas in Chapters II through IV.

Exercise 16. Let S be a finite subset of the integers. Prove that the functions $e^{im\theta}$ for $m \in S$ are linearly independent. Suggestion: Suppose some linear combination $v(\theta) = \sum c_m e^{im\theta}$ vanishes. Consider the integrals

$$\frac{1}{2\pi} \int_0^{2\pi} e^{-ik\theta} v(\theta) d\theta.$$

Our next result indicates the sense in which we may treat z and $\bar{z}$ as independent variables; the proof applies polar coordinates to demonstrate polarization!

Application I.7.2. (A simple case of polarization) Let $p : \mathbf{C} \times \mathbf{C} \to \mathbf{C}$ be a polynomial in two complex variables, and suppose $p(z, \bar{z}) = 0$ for all z. Then p is the zero polynomial; that is, $p(z, w) = 0$ for all z and w. As a consequence, if p and q satisfy $p(z, \bar{z}) = q(z, \bar{z})$ for all z, then $p(z, w) = q(z, w)$ for all z and w.

Proof. Assume $p(z, w) = \sum c_{jk} z^j w^k$. We will use polar coordinates to show that each coefficient c_{jk} vanishes. After setting $z = |z| e^{i\theta}$ and using the hypothesis we obtain

$$0 = p(|z| e^{i\theta}, |z| e^{-i\theta}) = \sum_{j,k} c_{jk} |z|^{j+k} e^{i(j-k)\theta}.$$

Changing indices with $m = j - k$ shows that

$$0 = \sum_m \sum_k c_{(k+m)k} |z|^{2k+m} e^{im\theta} = \sum_m C_m(|z|) e^{im\theta}.$$

By Exercise 16 the functions $e^{im\theta}$ are linearly independent, and therefore $C_m(|z|) = 0$ for all z. Therefore, after dividing by $|z|^m$, we have

$$0 = \sum_k c_{(k+m)k} |z|^{2k}$$

for all m and for all $z \neq 0$. Let p_m denote the polynomial defined by $p_m(t) = \sum_k c_{(k+m)k} t^k$. Each p_m vanishes at infinitely many points, and hence each p_m vanishes identically. Therefore the coefficients $c_{(k+m)k}$ all vanish, and p vanishes identically.   □

It is worthwhile to amplify the discussion about the independence of z and $\bar{z}$. Sometimes one thinks of z as a complex number; in this case $\bar{z}$ is its complex conjugate, and therefore it is determined by z. On the other hand, sometimes one thinks of z as a complex variable. There is a precise sense in which $\bar{z}$ is an *independent* variable. Application I.7.2 illustrates this by showing that we may vary these quantities separately in polynomial identities. The same holds more generally for identities in convergent power series in several complex variables, and a similar proof is valid. The process of separately varying the z and $\bar{z}$ variables is called *polarization* of an analytic identity. Here is a simple illustration. We may write the elementary identity

$$|z+1|^2 - |z-1|^2 = 4\mathrm{Re}(z)$$

in terms of z and $\bar{z}$:

$$(z+1)(\bar{z}+1) - (z-1)(\bar{z}-1) = 2(z+\bar{z}).$$

By Application I.7.2, we obtain, for every z and w, the identity

$$(z+1)(w+1) - (z-1)(w-1) = 2(z+w).$$

Let $p(z, w)$ define a polynomial in two (or $2n$) complex variables. As the notation suggests we write $p(z, \bar{z})$ for the value of $p(z, w)$ where w happens to be $\bar{z}$. Quite often in this book we will be studying functions of this type, while thinking only of the function $z \mapsto p(z, \overline{z})$. For such functions it makes sense by Application I.7.2 to view z and $\overline{z}$ as independent variables. We make a systematic study of these issues in Chapter VI.

We close this section by introducing without elaboration another viewpoint on polarization. Observe that the complex-valued functions z and $\overline{z}$ have linearly independent differentials dz and $d\overline{z}$. Therefore z and $\overline{z}$ are (locally) *functionally independent*. This independence justifies taking the partial derivatives of a smooth (infinitely differentiable) function with respect to both z and $\overline{z}$. Thus for example $\frac{\partial}{\partial z}(|z|^2) = \overline{z}$.

I.8   Roots of unity

The unit circle S^1 and the unit spheres in higher dimensions will arise many times throughout this book. We spend a few moments here to discuss some aspects of the unit circle. Let m be an integer with $m \geq 2$. A complex number ω is an m-th *root of unity* if $\omega^m = 1$. Then we have $|\omega| = 1$, so $\omega \in S^1$. A complex number ω is a *primitive m-th* root of unity if $\omega^m = 1$, but $\omega^k \neq 1$ for $1 \leq k < m$. For an m-th root of unity, the equation $\omega\overline{\omega} = 1$ implies

$$\overline{\omega} = \omega^{-1} = \omega^{m-1}.$$

Remark I.8.1. (For those familiar with groups) See the Appendix for the definition of a group. The unit circle is a group under multiplication. Let G be a finite cyclic group of order m. We may *represent* G as a subgroup of the unit circle. To do so, let g be a generator of G, and let ω be a primitive m-th root of unity. Define a function $\pi : G \to S^1$ by $\pi(g^j) = \omega^j$ for $1 \leq j \leq m$. Then the powers of g in G precisely correspond to the powers of ω in $\mathbf{C}$. Furthermore, π is a group homomorphism, because $\pi(g_1 g_2) = \pi(g_1)\pi(g_2)$ for all group elements g_1 and g_2.

The following simple identities about roots of unity are useful:

1) Let ω be an m-th root of unity. Then, if $\omega \neq 1$,

$$1 + \omega + \omega^2 + \cdots + \omega^{m-1} = 0. \tag{20}$$

Formula (20) follows immediately from formula (11) for the sum of a finite geometric series. It has a nice geometric interpretation. Think of ω as a point on S^1. When ω is a primitive m-th root of unity, its powers correspond to the vertices of a regular m-gon. If we think of these vertices as vectors based at the origin, then their vector sum must be zero by symmetry. A similar argument applies when ω is not a primitive m-th root of unity. See Exercise 17.

2) Let ω be a primitive m-th root of unity (thus $\omega \neq 1$), and assume $m \geq 3$. Then

$$\sum_{j=0}^{m-1} \omega^{2j} = 0.$$

This conclusion follows by replacing ω by ω^2 in (20). When $m = 2$, however, $\omega^2 = 1$, and substituting in (20) no longer is valid. In this case $1 + (-1)^2 = 2$.

Exercise 17. Provide a geometric proof of (20) in case $\omega^m = 1$, but ω is not a *primitive* m-th root of unity.

Exercise 18. Let z, ζ be complex numbers, and let ω be a primitive m-th root of unity, with $m \geq 3$. Evaluate

$$\frac{1}{m} \sum_{j=0}^{m-1} |z + \omega^j \zeta|^2 \omega^j.$$

Exercise 19.

1) Let ω be a primitive m-th root of unity. Evaluate

$$1 - \prod_{j=0}^{m-1} (1 - \omega^j x - \omega^j y).$$

2) (Difficult) Let ω be a primitive $(2r + 1)$-st root of unity. Evaluate

$$1 - \prod_{j=0}^{2r} (1 - \omega^j x - \omega^{2j} y).$$

The answer is a polynomial in x and y with positive integer coefficients. Try to find a formula for these integers. See [D] for more information.

I.9   Summary

We summarize this introductory chapter. Our starting point was the existence of the real number system $\mathbf{R}$ and the elementary properties of a complete ordered field. We have defined the complex number system $\mathbf{C}$ in terms of $\mathbf{R}$ and shown it to be a complete field. The complex numbers do not form an *ordered* field. We have therefore considered inequalities about absolute values of complex numbers. Doing so has enabled us both to provide geometric interpretations of operations with complex numbers and to set down the basic principles for this book. We have introduced both analytic notions such as convergent power series and algebraic notions such as roots of unity. Many of the computations in this chapter hold in greater generality, and thus the chapter has prepared us for deeper discussion.

CHAPTER II

Complex Euclidean Spaces and Hilbert Spaces

Next we extend the ideas we found useful in $\mathbf{C}$ to both finite-dimensional and infinite-dimensional vector spaces over $\mathbf{C}$. See Appendix A.1 for the list of axioms for a vector space. Most of the geometric reasoning from Chapter I applies in the setting of Hilbert spaces. One of the main new issues will be the existence of orthogonal projections onto closed subspaces. This notion is related to a version of the Pythagorean theorem. We introduce bounded linear transformations (operators) between Hilbert spaces in this chapter, but postpone results about adjoints and Hermitian forms to Chapter IV.

The Riesz representation lemma characterizing bounded linear functionals is an important consequence of orthogonal projection; it appears in Section II. Orthonormal expansion and polarization identities also appear in this chapter. Finally we exhibit some orthonormal systems arising in physics and engineering, and interpret them from the point of view of generating functions. This discussion anticipates one aspect of the proof of Theorem VII.1.1.

II.1 Hermitian inner products

We saw in Chapter I that Euclidean geometry arises from the notion of the absolute value of a complex number. The ideas of Euclidean geometry extend to Hilbert spaces. A Hilbert space is a complex vector space with a Hermitian inner product and corresponding norm making it into a complete metric space. We begin by discussing the Euclidean inner product and Euclidean norm in finite-dimensional complex vector spaces.

We let $\mathbf{C}^n$ denote complex Euclidean space of dimension n. As a set, $\mathbf{C}^n$ consists of all n-tuples of complex numbers; we write $z = (z_1, \ldots, z_n)$ for a point in $\mathbf{C}^n$. This set has the structure of a complex vector space with the usual operations of vector addition and scalar multiplication, so we sometimes say *vector* instead of point in $\mathbf{C}^n$. We compute the Euclidean inner product of two vectors by formula (1) below. The notation $\mathbf{C}^n$ includes the vector space structure, the Hermitian inner product defined by (1), and the squared norm defined by (2).

The *Euclidean inner product* is given by

$$\langle z, w \rangle = \sum_{j=1}^{n} z_j \overline{w}_j \tag{1}$$

and the *Euclidean squared norm* is given by

$$\| z \|^2 = \langle z, z \rangle. \tag{2}$$

In general a *norm* on a complex vector space V is a function $v \mapsto \| v \|$ satisfying the following three properties:

N1) $\| v \| > 0$ for all nonzero v.

N2) $\| cv \| = | c | \, \| v \|$ for all $c \in \mathbf{C}$ and all $v \in V$.

N3) (The triangle inequality) $\| v + w \| \leq \| v \| + \| w \|$ for all $v, w \in V$.

The Euclidean norm on $\mathbf{C}^n$ is a norm in this sense. Properties N1) and N2) are evident. The proof of N3) is virtually the same as the proof of (10) from Chapter I. The same proof works in any complex vector space with a norm arising from a Hermitian inner product. We therefore give the appropriate definitions and then prove the triangle inequality in that generality.

Definition II.1.1. (**Hermitian inner product**) Let V be a complex vector space. A *Hermitian inner product* on V is a function $\langle \, , \, \rangle$ from $V \times V$ to $\mathbf{C}$ satisfying the following four properties. For all u, v, $w \in V$, and for all $c \in \mathbf{C}$:

1) $\langle u + v, w \rangle = \langle u, w \rangle + \langle v, w \rangle$.
2) $\langle cu, v \rangle = c \langle u, v \rangle$.
3) $\langle u, v \rangle = \overline{\langle v, u \rangle}$.
4) $\langle u, u \rangle > 0$ for $u \neq 0$.

Three additional properties are consequences:

5) $\langle u, v + w \rangle = \langle u, v \rangle + \langle u, w \rangle$.
6) $\langle u, cv \rangle = \overline{c} \langle u, v \rangle$.
7) $\langle 0, w \rangle = 0$ for all $w \in \mathcal{H}$. In particular $\langle 0, 0 \rangle = 0$.

A Hermitian inner product is often called a positive definite sesquilinear form. The prefix "sesqui" means one and a half times; for example a sesquicentennial is a 150 year celebration. Since the form $\langle u, v \rangle$ is linear in u and conjugate linear in v, and therefore not "bilinear," sesquilinear seems an appropriate term. Property 4) is *positive definiteness*; its consequences are crucial for the next definition and much of what follows.

Positive definiteness provides a technique for verifying that a given z equals 0. It follows from 4) and 7) that $z = 0$ if and only if $\langle z, w \rangle = 0$ for all w in $\mathcal{H}$.

Definition II.1.2. The *norm* $\| \ \|$ corresponding to the Hermitian inner product $\langle \, , \, \rangle$ is defined by

$$\| v \| = \sqrt{\langle v, v \rangle}.$$

Theorem II.1.3 below justifies the terminology. The distance function corresponding to $\langle \, , \, \rangle$ (or to $\| \ \|$) is given by

$$d(u, v) = \| u - v \|.$$

The function d is symmetric in its arguments u and v, its values are nonnegative, and its values are positive when $u \neq v$. To show that it is a distance function in the metric space sense (Definition I.5.1), we need also to establish the triangle inequality

$$\| u - \zeta \| \leq \| u - v \| + \| v - \zeta \|.$$

As before this follows immediately from the triangle inequality for the norm:

$$\| z + w \| \leq \| z \| + \| w \|.$$

Theorem II.1.3. (**The Cauchy-Schwarz and triangle inequalities**) Let V be a complex vector space, let $\langle \ , \ \rangle$ be a Hermitian inner product on V, and let $\| v \| = \sqrt{\langle v, v \rangle}$. The following inequalities are valid for all $z, w \in V$, and $\| \ \|$ is a norm on V:

$$| \langle z, w \rangle | \leq \| z \| \| w \| \tag{3}$$

$$\| z + w \| \leq \| z \| + \| w \|. \tag{4}$$

Proof. Properties N1) and N2) of a norm are evident; N1) follows from property 4) of the inner product, and N2) follows by properties 2) and 6) of the inner product. Property N3) is the triangle inequality (4).

We first prove the Cauchy-Schwarz inequality (3). The reader should compare this proof with Application I.3.1. For all $t \in \mathbf{C}$, and for all z and w in V,

$$0 \leq \| z + tw \|^2 = \| z \|^2 + 2\mathrm{Re}\langle z, tw \rangle + | t |^2 \| w \|^2. \tag{5}$$

Think of z and w as fixed, and let ϕ be the quadratic polynomial in t and $\bar{t}$ defined by the right-hand side of (5). The values of ϕ are nonnegative; we seek its minimum value by setting its differential equal to 0. We use subscripts to denote the derivatives with respect to t and $\bar{t}$. (See the discussion after Application I.7.2.) Since ϕ is real-valued, we have $\phi_t = 0$ if and only if $\phi_{\bar{t}} = 0$. From (5) we find

$$\phi_{\bar{t}} = \langle z, w \rangle + t \| w \|^2.$$

When $w = 0$ we get no useful information, but the desired inequality (3) is true when $w = 0$. To prove (3) when $w \neq 0$, we may set

$$t = \frac{-\langle z, w \rangle}{\| w \|^2}$$

in (5) and conclude that

$$0 \leq \| z \|^2 - 2 \frac{|\langle z, w \rangle|^2}{\| w \|^2} + \frac{|\langle z, w \rangle|^2}{\| w \|^2} = \| z \|^2 - \frac{|\langle z, w \rangle|^2}{\| w \|^2}. \quad (6)$$

Inequality (6) yields

$$|\langle z, w \rangle|^2 \leq \| z \|^2 \| w \|^2,$$

from which the Cauchy-Schwarz inequality follows by taking square roots.

To establish the triangle inequality (4), we begin by squaring its left-hand side:

$$\| z + w \|^2 = \| z \|^2 + 2\mathrm{Re}\langle z, w \rangle + \| w \|^2.$$

Since $\mathrm{Re}\langle z, w \rangle \leq |\langle z, w \rangle|$, the Cauchy-Schwarz inequality yields

$$\begin{aligned} \| z + w \|^2 &= \| z \|^2 + 2\mathrm{Re}\langle z, w \rangle + \| w \|^2 \\ &\leq \| z \|^2 + 2\| z \| \, \| w \| + \| w \|^2 \\ &= (\| z \| + \| w \|)^2. \end{aligned}$$

Taking the square root of each side gives the triangle inequality and completes the proof that $\sqrt{\langle v, v \rangle}$ defines a norm on V. $\square$

The two inequalities from Theorem II.1.3 have many consequences. For example, we use them to show that the inner product and norm on V are (sequentially) continuous functions. Let $\{z_n\}$ be a sequence in V. By definition $\{z_n\}$ converges to z if the sequence of real numbers $\| z_n - z \|$ converges to 0. We have the following result:

Proposition II.1.4. (Continuity of the inner product and the norm) Let V be a complex vector space with Hermitian inner product and corresponding norm. Suppose that $\{z_n\}$ is a sequence that converges to z. Then, for all $w \in V$, the sequence of inner products $\langle z_n, w \rangle$ converges to $\langle z, w \rangle$. Furthermore $\|z_n\|$ converges to $\|z\|$.

Proof. The first statement follows from the linearity of the inner product and the Cauchy-Schwarz inequality. We have

$$|\langle z_n, w \rangle - \langle z, w \rangle| = |\langle z_n - z, w \rangle| \le \|z_n - z\| \, \|w\|.$$

Thus, when z_n converges to z, the right-hand side converges to 0, and therefore so does the left-hand side. The inner product of the limit is thus the limit of the inner products.

The proof of the second statement uses the triangle inequality. From it we obtain the inequality $\|z\| \le \|z - z_n\| + \|z_n\|$ and hence

$$\|z\| - \|z_n\| \le \|z - z_n\|.$$

Interchanging the roles of z_n and z gives the same inequality with a negative sign on the left-hand side. Combining these inequalities yields

$$|\, \|z\| - \|z_n\| \,| \le \|z - z_n\|,$$

from which the desired statement follows. $\qquad\qquad\qquad\qquad$ $\square$

    One important consequence of Proposition II.1.4 concerns the interchange of limit and summation. Suppose that $\sum v_n$ converges in $\mathcal{H}$. For all $w \in \mathcal{H}$, we have

$$\left\langle \sum_n v_n, w \right\rangle = \sum_n \langle v_n, w \rangle.$$

This conclusion follows immediately from applying Proposition II.1.4 to the partial sums of the series. When working with orthonormal expansions we will often apply this result without comment.

The inequalities from Theorem II.1.3 also lie behind complex geometry. For example, the formula for the squared norm of a sum generalizes the *law of cosines* from elementary trigonometry, and the particular case when the vectors are orthogonal generalizes the Pythagorean theorem. In complex Euclidean space we therefore have the usual notions of angles, lines, and balls. In particular the *open unit ball* in $\mathbf{C}^n$ will play a significant role, so we use special notation for it:

$$\mathbf{B}_n = \{z \in \mathbf{C}^n : \|z\| < 1\}.$$

For emphasis we note that $\mathbf{B}_n$ is the *open* unit ball. Its boundary is the unit sphere S^{2n-1}; the number $2n - 1$ is the *real* dimension of the unit sphere in $\mathbf{C}^n$.

Exercise 1. The inequality $|z - w| \leq |1 - z\overline{w}|$ holds for all complex numbers z and w in the closed unit disk, by Exercise 3 of Chapter I.

1) Show that the inequality

$$\|z - w\| \leq |1 - \langle z, w \rangle|$$

may fail for z and w in the closed unit ball in two or more dimensions.

2) For z, w in the closed unit ball, prove the inequality

$$\|z - w\|^2 \leq 2|1 - \langle z, w \rangle|.$$

3) Determine all positive values of c and α for which the inequality

$$\|z - w\| \leq c|1 - \langle z, w \rangle|^{\alpha}$$

holds for all z, w in the closed unit ball in $\mathbf{C}^n$.

Although our primary interest lies in properties of complex Euclidean space, the proofs of many of our results involve infinite-dimensional spaces of functions. Hilbert spaces are those normed complex vector spaces whose geometric properties most closely resemble those of complex Euclidean spaces.

Definition II.1.5. A *Hilbert space* $\mathcal{H}$ is a complex vector space, together with a Hermitian inner product whose corresponding distance function makes $\mathcal{H}$ into a complete metric space.

The norm on a Hilbert space arises from the Hermitian inner product. In many situations one works with complete normed vector spaces whose norm does not arise from an inner product. A *Banach space* V is a real or complex vector space together with a norm $\|\ \|$ whose corresponding distance function makes V into a complete metric space. Several times in this book we will mention some of the standard Banach spaces, so we pause now to define them. See [F] for details.

Consider the space $C(X)$ of continuous complex-valued functions on a compact metric space X. We make $C(X)$ into a normed linear space by using the supremum norm; this norm is defined by $\|f\|_\infty = \sup_{x \in X} |f(x)|$. Convergence in $C(X)$ means *uniform convergence*. Since the uniform limit of continuous functions is continuous, $C(X)$ is complete. It is therefore a Banach space, but (unless X is a finite set) there is no way to make $C(X)$ into a Hilbert space.

Next let X be a nice (such as open or closed) subset of Euclidean space $\mathbf{R}^n$, let dV denote Lebesgue measure, and suppose $1 \leq p < \infty$. The spaces $L^p(X)$ consist of (equivalence classes of) measurable functions f for which $\int_X |f(t)|^p \, dV(t) < \infty$. We define the L^p norm by

$$\|f\|_p = \left(\int_X |f(x)|^p \, dV(x) \right)^{\frac{1}{p}}.$$

One of the principal results of the Lebesgue theory of integration is the completeness of the L^p spaces. See [F]. Thus $L^p(X)$ is a Banach space; it is a Hilbert space only when $p = 2$. As mentioned above, we will work mostly within the context of Hilbert spaces; therefore we will drop the subscript from the L^2 norm unless it is required for clarity.

Exercise 2. Let $\mathcal{H}$ be a Hilbert space. Show that, for all z and w in $\mathcal{H}$,

$$|| z + w ||^2 + || z - w ||^2 = 2|| z ||^2 + 2|| w ||^2, \tag{7}$$

and give a geometric interpretation.

Exercise 3. (Difficult) Suppose that V is a complex Banach space whose norm satisfies (7). Prove that there is an inner product making V into a Hilbert space. Suggestion: Use the polarization identity (16) from Section II.4 below when $m = 4$ to determine a candidate for the inner product of two elements. Verify that this candidate satisfies the required properties. To prove the multiplicative property $\langle cz, w \rangle = c\langle z, w \rangle$, first prove it by induction for c a positive integer. Then prove it for c rational, then for c real, and finally for c complex.

The following examples indicate the wide applicability of Hilbert spaces. We elaborate some of them in later parts of the book. Examples H4) and H5) from Example II.1.6 require the notion of holomorphic function of several complex variables; this topic appears in Chapter III. In this book the two most important examples of Hilbert spaces are complex Euclidean space $\mathbf{C}^n$ and H4) from Example II.1.6, when the domain Ω there is the unit ball.

Example II.1.6. (Hilbert Spaces)

H1) Finite-dimensional complex Euclidean space $\mathbf{C}^n$ is a Hilbert space. In particular $\mathbf{C}^n$ is a complete metric space with the distance function given by $d(z, w) = || z - w ||$. See Exercise 4.

H2) l^2. Let $a = \{a_\nu\}$ denote a sequence of complex numbers. We say that a is square-summable, and we write $a \in l^2$, if $\sum_\nu |a_\nu|^2$ is finite. When $a, b \in l^2$ we write

$$\langle a, b \rangle = \sum_\nu a_\nu \overline{b}_\nu$$

for their Hermitian inner product. Exercise 4 requests a proof that l^2 is a complete metric space; the distance function is the one associated with this Hermitian inner product.

H3) $L^2(\Omega)$. Let Ω be an open subset of $\mathbf{R}^n$. Let dV denote Lebesgue measure in $\mathbf{R}^n$. We write $L^2(\Omega)$ for the complex vector space of (equivalence classes of) measurable functions $f : \Omega \to \mathbf{C}$ for which $\int_{\Omega} |f(x)|^2 \, dV(x)$ is finite. When f and g are elements of $L^2(\Omega)$ we define their inner product by

$$\langle f, g \rangle = \int_{\Omega} f(x)\overline{g(x)} \, dV(x).$$

The corresponding norm and distance function make $L^2(\Omega)$ into a complete metric space, so $L^2(\Omega)$ is a Hilbert space. See [F] for a proof of completeness.

Recall that a *linear subspace* of a vector space is a subset that is itself a vector space under the same operations of addition and scalar multiplication. One often omits the word "linear" when the context makes it clear. A finite-dimensional subspace of a Hilbert space is necessarily closed (in the metric space sense) whereas infinite-dimensional subspaces need not be closed. A closed linear subspace of a Hilbert space is complete, and hence is also a Hilbert space. Here is our favorite example.

H4) Let Ω be a bounded open subset of $\mathbf{C}^n$. Let $A^2(\Omega)$ denote the space of holomorphic functions f on Ω such that $f \in L^2(\Omega)$. In Corollary III.1.8 we show that $A^2(\Omega)$ is a closed subspace of $L^2(\Omega)$, and hence it is also complete. This example is particularly important in complex analysis. The linear transformation that projects $L^2(\Omega)$ to $A^2(\Omega)$ is called the Bergman projection and will play a crucial role in this book.

H5) Let $A^2(\mathbf{C}^n, dG)$ consist of those holomorphic functions f on $\mathbf{C}^n$ such that

$$\| f \|_G^2 = \int |f(z)|^2 e^{-\|z\|^2} \, dV(z) < \infty.$$

Then $A^2(\mathbf{C}^n, dG)$ is a Hilbert space. The letter G stands for *Gaussian*, and dG is the measure given by $e^{-\|z\|^2} dV(z)$.

Exercise 4. Verify that every Cauchy sequence in $\mathbf{C}^n$ converges. Prove that l^2 is complete.

We conclude this section by introducing *bounded linear transformations* or *operators*. These are the continuous functions between Hilbert spaces that preserve the vector space structure. We provide a systematic treatment in Chapter IV.

Definition II.1.7. Let $\mathcal{H}$ and $\mathcal{H}'$ be Hilbert spaces. A function $L : \mathcal{H} \to \mathcal{H}'$ is called a *bounded linear transformation* from $\mathcal{H}$ to $\mathcal{H}'$ if L satisfies the following three properties:

1) $L(z_1 + z_2) = L(z_1) + L(z_2)$ for all z_1 and z_2 in $\mathcal{H}$.

2) $L(cz) = cL(z)$ for all $z \in \mathcal{H}$ and all $c \in \mathbf{C}$.

3) There is a constant C such that $\| L(z) \| \leq C \|z\|$ for all $z \in \mathcal{H}$.

Properties 1) and 2) are the *linearity* of L. Property 3) guarantees the continuity of L; see Lemma II.1.9 below. The infimum of the set of constants C that work in 3) provides a measurement of the size of the transformation L; it is called the norm of L, and is written $\| L \|$. An equivalent way to define $\| L \|$ is by the formula

$$\| L \| = \sup_{\{z \neq 0\}} \frac{\| L(z) \|}{\| z \|}.$$

The notation $\| \ \|$ has three different meanings, but the context prevents ambiguity. This notation is used for the norm on the domain $\mathcal{H}$, for the norm on the range $\mathcal{H}'$, and for a norm on the Banach space (See Theorem V.1.4) of linear transformations between them.

The special case where the target $\mathcal{H}'$ of L is $\mathbf{C}$ is especially important; we will study this situation in Section 2.

Definition II.1.8. A *bounded linear functional* on a Hilbert space $\mathcal{H}$ is a bounded linear transformation from $\mathcal{H}$ to $\mathbf{C}$.

We next discuss the relationship between boundedness and continuity for linear transformations.

Lemma II.1.9. Let $L : \mathcal{H} \to \mathcal{H}'$ be a linear transformation between Hilbert spaces. The following three statements are equivalent:

1) There is a constant $C > 0$ such that, for all z,

$$\| Lz \| \le C \| z \|.$$

2) L is continuous at the origin.

3) L is continuous at every point.

Proof. It follows from the ϵ-δ definition of continuity at a point and the linearity of L that statements 1) and 2) are equivalent. See Exercise 5. Certainly statement 3) implies statement 2). Finally statement 1) and the linearity of L imply statement 3) because

$$\| Lz - Lw \| = \| L(z - w) \| \le C \| z - w \|. \qquad \square$$

Exercise 5. Consider the proof that 2) implies 1) from Lemma II.1.9. For a given ϵ, suppose δ works in the ϵ-δ definition of continuity. What values of C work in 1)?

Observe that a bounded linear transformation is defined everywhere. Linear transformations whose natural domain of definition is a dense subspace of a Hilbert space often arise in mathematics and physics. Suppose that V is a dense subspace of $\mathcal{H}$ and that $L : V \to \mathcal{H}'$ is a linear transformation on V. There are two possibilities.

1) L extends to be a bounded linear transformation from $\mathcal{H}$ to $\mathcal{H}'$; that is, there is a linear transformation whose restriction to V agrees with L.

2) There is no such linear transformation.

When 2) applies it is common to say that L is an *unbounded linear transformation* or *unbounded operator* from $\mathcal{H}$ to $\mathcal{H}'$, even though it is not defined everywhere. This situation requires careful consideration of what is the domain of L.

We give a simple example to close this section. A function on **R** has compact support if it vanishes outside of some compact set. Suppose that $\mathcal{H} = L^2(\mathbf{R})$ and that V is the dense subspace of infinitely differentiable functions with compact support. Define L by $L(f) = f'$, where f' denotes the usual derivative of f. Then L is linear on V, but L is not the restriction to V of a linear transformation on $\mathcal{H}$.

II.2 Orthogonality, projections and closed subspaces

Let $\mathcal{H}$ be a Hilbert space, and suppose $z, w \in \mathcal{H}$. We say that z and w are *orthogonal* if $\langle z, w \rangle = 0$. Orthogonality generalizes perpendicularity and thereby provides geometric insight in the general Hilbert space setting. The term "orthogonal" applies also for subspaces. Subspaces V and W of $\mathcal{H}$ are orthogonal if $\langle v, w \rangle = 0$ for all $v \in V$ and $w \in W$. We say that z is orthogonal to V if $\langle z, v \rangle = 0$ for all v in V, or equivalently, if the one-dimensional subspace generated by z is orthogonal to V.

Suppose that V and W are orthogonal closed subspaces of a Hilbert space. We write $V \oplus W$ for their orthogonal sum; it is the subspace of $\mathcal{H}$ consisting of those z that can be written $z = v + w$, where $v \in V$ and $w \in W$. We sometimes write $z = v \oplus w$ in order to emphasize orthogonality. Then $\| v \oplus w \|^2 = \| v \|^2 + \| w \|^2$; this is the Pythagorean Theorem. An immediate consequence of the Pythagorean theorem is that $v \oplus w = 0$ if and only if both $v = 0$ and $w = 0$.

We now study the geometric notion of orthogonal projection onto a closed subspace. The next theorem guarantees that we can *project* a vector w in a Hilbert space onto a closed subspace. This existence and uniqueness theorem has diverse corollaries.

Theorem II.2.1. Let V be a closed subspace of a Hilbert space $\mathcal{H}$. For each w in $\mathcal{H}$ there is a unique $z \in V$ that minimizes $\| z - w \|$. This z is the *projection* of w onto V.

Proof. Fix w. If $w \in V$ then the conclusion holds with $z = w$. In general let $d = \inf_{z \in V} \| z - w \|$. Choose a sequence $\{z_n\}$ such that $z_n \in V$ for all n and $\| z_n - w \|$ tends to d. We will show that $\{z_n\}$ is a Cauchy sequence, and hence it converges to some z; by continuity of the norm (Proposition II.1.4) we then have $\| z - w \| = d$.

Following the geometric insight provided by Application I.3.4, we use the parallelogram law (Exercise 2) to express $\| z_n - z_m \|^2$ as follows:

$$\| z_n - z_m \|^2 = \| (z_n - w) + (w - z_m) \|^2 = 2\| z_n - w \|^2$$
$$+ 2\| w - z_m \|^2 - \| (z_n - w) - (w - z_m) \|^2.$$

The last term on the right-hand side is

$$4 \left\| \frac{z_n + z_m}{2} - w \right\|^2.$$

Since V is a subspace, the midpoint $\frac{z_n + z_m}{2}$ lies in V as well. Therefore this term is at least $4d^2$, and we obtain

$$0 \le \| z_n - z_m \|^2 \le 2\| z_n - w \|^2 + 2\| w - z_m \|^2 - 4d^2. \quad (8)$$

As m and n tend to infinity the right-hand side of (8) tends to $2d^2 + 2d^2 - 4d^2 = 0$. Thus $\{z_n\}$ is a Cauchy sequence in $\mathcal{H}$ and hence converges. Since V is closed, the limit is in V.

It remains only to show uniqueness. The proof precisely parallels the proof of Corollary I.3.5. Given a pair of minimizers z and ζ, let d_m^2 denote the squared distance from their midpoint to w. By the parallelogram law we may write

$$2d^2 = ||z - w||^2 + ||\zeta - w||^2 = 2\left\|\frac{z + \zeta}{2} - w\right\|^2 + 2\left\|\frac{z - \zeta}{2}\right\|^2$$

$$= 2d_m^2 + 2\left\|\frac{z - \zeta}{2}\right\|^2.$$

Thus $d^2 \geq d_m^2$, and strict inequality holds unless $z = \zeta$. Thus the minimizer is unique. $\qquad\square$

Corollary II.2.2. Let V be a closed subspace of a Hilbert space $\mathcal{H}$. For each $w \in \mathcal{H}$ there is a unique way to write $w = v + \zeta = v \oplus \zeta$ where $v \in V$ and ζ is orthogonal to V.

Proof. Let v be the projection of w onto V guaranteed by Theorem II.2.1. Since $w = v + (w - v)$, the existence result follows if we can show that $w - v$ is orthogonal to V. To see the orthogonality choose $u \in V$. Then consider the function f of one complex variable defined by

$$f(\lambda) = ||v + \lambda u - w||^2.$$

By Theorem II.2.1, f achieves its minimum at $\lambda = 0$. Therefore for all λ

$$0 \leq f(\lambda) - f(0) = 2\operatorname{Re}\langle v - w, \lambda u \rangle + |\lambda|^2 ||u||^2. \qquad (9)$$

We claim that (9) forces $\langle v - w, u \rangle = 0$. Granted the claim, we note that u is an arbitrary element of V. Therefore $v - w$ is orthogonal to V, as required.

There are many ways to prove the claim, thereby completing the proof of existence. One way is to note that $\langle v - w, u \rangle$ is the (partial) derivative of f with respect to $\bar{\lambda}$ at 0, and hence vanishes at a minimum of f. A second method is to consider the inequality first when λ is real, and then again when λ is imaginary. It follows that both the real and imaginary parts of $\langle v - w, u \rangle$ must vanish.

The uniqueness assertion is easy; we use the notation for orthogonal sum. Suppose $w = v \oplus \zeta = v' \oplus \zeta'$ as in the statement of the

Corollary. Then

$$0 = w - w = (v - v') \oplus (\zeta - \zeta')$$

from which we obtain $v = v'$ and $\zeta = \zeta'$. □

Corollary II.2.3. Let V be a closed subspace of a Hilbert space $\mathcal{H}$. For each $w \in \mathcal{H}$, let Pw denote the unique $z \in V$ guaranteed by Theorem II.2.1; Pw is also the v guaranteed by Corollary II.2.2. Then the mapping $w \to P(w)$ is a bounded linear transformation satisfying $P^2 = P$.

Proof. Both the existence and uniqueness assertions in Corollary II.2.2 are crucial. Given w_1 and w_2 in $\mathcal{H}$, by existence we may write $w_1 = Pw_1 \oplus \zeta_1$ and $w_2 = Pw_2 \oplus \zeta_2$. Adding these gives

$$w_1 + w_2 = (Pw_1 \oplus \zeta_1) + (Pw_2 \oplus \zeta_2)$$
$$= (Pw_1 + Pw_2) \oplus (\zeta_1 + \zeta_2). \tag{10}$$

The uniqueness assertion and (10) show that $Pw_1 + Pw_2$ is the unique element of V corresponding to $w_1 + w_2$ guaranteed by Corollary II.2.2; by definition this element is $P(w_1 + w_2)$. By uniqueness $Pw_1 + Pw_2 = P(w_1 + w_2)$, and P is additive. In a similar fashion we write $w = Pw \oplus \zeta$ and hence

$$cw = c(Pw) \oplus c\zeta.$$

Again by uniqueness, $c(Pw)$ must be the unique element corresponding to cw guaranteed by Corollary II.2.2; by definition this element is $P(cw)$. Hence $cP(w) = P(cw)$. We have now shown that P is linear.

To show that P is bounded, we note from the Pythagorean theorem that $|| w ||^2 = || Pw ||^2 + || \zeta ||^2$. This equality implies the inequality $|| Pw || \leq || w ||$.

Finally we show that $P^2 = P$. For $z = v \oplus \zeta$, we have $P(z) = v = v \oplus 0$. Hence

$$P^2(z) = P(P(z)) = P(v \oplus 0) = v = P(z). \qquad □$$

Theorem II.2.1 and its consequences are among the most powerful results in the book. The theorem guarantees that we can solve a minimization problem in diverse infinite-dimensional settings, and it implies the Riesz representation lemma.

Fix $w \in \mathcal{H}$, and consider the function from $\mathcal{H}$ to $\mathbf{C}$ defined by $Lz = \langle z, w \rangle$. Then L is a bounded linear functional. The linearity is evident. The boundedness follows from the Cauchy-Schwarz inequality; setting $C = \|w\|$ guarantees that $|L(z)| \leq C\|z\|$ for all $z \in \mathcal{H}$.

The following fundamental result of F. Riesz characterizes bounded linear functionals on a Hilbert space; a bounded linear functional must be given by an inner product. The proof relies on projection onto a closed subspace.

Theorem II.2.4. (Riesz Lemma) Let $\mathcal{H}$ be a Hilbert space and suppose that L is a bounded linear functional on $\mathcal{H}$. Then there is a unique $w \in \mathcal{H}$ such that

$$L(z) = \langle z, w \rangle$$

for all $z \in \mathcal{H}$. Furthermore, the norm $\|L\|$ of the linear transformation L equals $\|w\|$.

Proof. Let $N(L)$ denote the nullspace of L, namely the set of z such that $L(z) = 0$. Since L is linear, $N(L)$ is a subspace of $\mathcal{H}$. Since L is bounded, it is continuous, and so $N(L)$ is closed. If $N(L) = \mathcal{H}$, we take $w = 0$ and the result is true.

Suppose that $N(L)$ is not $\mathcal{H}$. Theorem II.2.1 implies that there is a nonzero element w_0 orthogonal to $N(L)$. To find such a w_0, choose any nonzero element not in $N(L)$ and subtract its orthogonal projection onto $N(L)$.

Let now z be an arbitrary element of $\mathcal{H}$. For a complex number α we can write

$$z = (z - \alpha w_0) + \alpha w_0.$$

Note that $L(z - \alpha w_0) = 0$ if and only if $\alpha = \frac{L(z)}{L(w_0)}$. For each z we therefore let $\alpha_z = \frac{L(z)}{L(w_0)}$.

Since w_0 is orthogonal to $N(L)$, computing the inner product with w_0 yields

$$\langle z, w_0 \rangle = \alpha_z \| w_0 \|^2 = \frac{L(z)}{L(w_0)} \| w_0 \|^2. \tag{11}$$

From (11) we see that

$$L(z) = \left\langle z, \frac{w_0}{\| w_0 \|^2} \overline{L(w_0)} \right\rangle$$

and the existence result is proved.

The reader should note the formula for w:

$$w = \frac{w_0}{\| w_0 \|^2} \overline{L(w_0)}.$$

The uniqueness for w is immediate from the test we mentioned earlier. If $\langle \zeta, w - w' \rangle$ vanishes for all ζ, then $w - w' = 0$.

It remains to show that $\| L \| = \| w \|$. The Cauchy-Schwarz inequality yields

$$\| L \| = \sup_{\| z \| = 1} | \langle z, w \rangle | \leq \| w \|.$$

Choosing $\frac{w}{\| w \|}$ for z yields

$$\| L \| \geq \left| L \left(\frac{w}{\| w \|} \right) \right| = \frac{\langle w, w \rangle}{\| w \|} = \| w \|.$$

Combining the two inequalities shows that $\| L \| = \| w \|$. $\qquad\square$

Exercise 6. A hyperplane in $\mathcal{H}$ is a level set of a non-trivial linear functional. Assume that $w \neq 0$. Find the distance between the parallel hyperplanes given by $\langle z, w \rangle = c_1$ and $\langle z, w \rangle = c_2$.

Exercise 7. Let $b = \{b_j\}$ be a sequence of complex numbers, and suppose there is a positive number C such that

$$\left| \sum_{j=1}^{\infty} a_j \bar{b}_j \right| \leq C \left(\sum_{j=1}^{\infty} |a_j|^2 \right)^{\frac{1}{2}}$$

for all $a \in l^2$. Show that $b \in l^2$ and that $\sum |b_j|^2 \leq C^2$. Suggestion: Consider the map that sends a to $\sum a_j \bar{b}_j$.

II.3   Orthonormal expansion

We continue our general discussion of the theory of Hilbert spaces by discussing orthonormal expansions. For the rest of this book we assume that a Hilbert space is *separable*. This term means that the Hilbert space has a countable dense set; separability implies that the orthonormal systems we are about to define are either finite or countably infinite sets. All the specific Hilbert spaces mentioned or used in this book are separable. Some of the proofs given tacitly use separability even when the result holds more generally.

Definition II.3.1. Let $S = \{z_n\}$ be a finite or countably infinite collection of elements in a Hilbert space $\mathcal{H}$. We say that S is an *orthonormal system* in $\mathcal{H}$ if, for each n we have $\| z_n \|^2 = 1$, and for each n, m with $n \neq m$, we have $\langle z_n, z_m \rangle = 0$. We say that S is a *complete orthonormal system* if, in addition, $\langle z, z_n \rangle = 0$ for all n implies $z = 0$.

Suppose that $S = \{z_n\}$ is a countably infinite orthonormal system in $\mathcal{H}$. Choose $z \in \mathcal{H}$. For each positive integer N we obtain inequality (12) by using orthonormality:

$$0 \leq \| z - \sum_{n=1}^{N} \langle z, z_n \rangle z_n \|^2 = \| z \|^2 - \sum_{n=1}^{N} |\langle z, z_n \rangle|^2. \qquad (12)$$

Thus the sequence of real numbers $r_N = r_N(z)$ given by

$$r_N = \sum_{n=1}^{N} |\langle z, z_n \rangle|^2$$

is bounded above and nondecreasing. Therefore it has a limit $r = r(z)$.
By (12) we obtain Bessel's inequality

$$r(z) = \sum_{n=1}^{\infty} | \langle z, z_n \rangle |^2 \le \| z \|^2. \tag{13}$$

The limit $r(z)$ will equal $\| z \|^2$ for each z if and only if the orthonormal system S is *complete*. This statement is the content of the following fundamental theorem. In general $r(z)$ is the squared norm of the projection of z onto the span of the z_j.

Theorem II.3.2. **(Orthonormal Expansion)** An orthonormal system $S = \{z_n\}$ is complete if and only if, for each $z \in \mathcal{H}$, we have

$$z = \sum_n \langle z, z_n \rangle z_n. \tag{14}$$

Proof. The cases where S is a finite set or where $\mathcal{H}$ is finite-dimensional are evident. So we assume that $\mathcal{H}$ is infinite-dimensional and that S is a countably infinite set. We first verify that the sum on the right-hand side in (14) makes sense. Fix $z \in \mathcal{H}$. Put

$$T_N = \sum_{n=1}^{N} \langle z, z_n \rangle z_n$$

and, for $N > M$, observe that

$$\| T_N - T_M \|^2 = \| \sum_{n=M+1}^{N} \langle z, z_n \rangle z_n \|^2$$

$$= \sum_{n=M+1}^{N} | \langle z, z_n \rangle |^2 = r_N - r_M. \tag{15}$$

Since $\{r_N\}$ converges, it is a Cauchy sequence of real numbers. Thus (15) implies that $\{T_N\}$ is a Cauchy sequence in $\mathcal{H}$. Since $\mathcal{H}$ is complete, T_N converges to some element w of $\mathcal{H}$, and $w = \sum \langle z, z_n \rangle z_n$. Note that $\langle w, z_n \rangle = \langle z, z_n \rangle$ for each n, so $z - w$ is orthogonal to each z_n.

We can now establish both implications. Suppose first that S is a complete system. Since $z - w$ is orthogonal to each z_n, we have $z - w = 0$. Thus (14) holds. Conversely, suppose that (14) holds. To show that S is a complete system, we assume that $\langle z, z_n \rangle = 0$ for all n, and hope to show that $z = 0$. This conclusion follows immediately from (14). $\qquad\square$

Exercise 8. Show that the functions $e^{2\pi i n x}$, as n ranges over all integers, form a complete orthonormal system for $L^2[0, 1]$. Suggestion: To verify completeness, first show that smooth (infinitely differentiable) periodic functions are dense.

II.4   The polarization identity

Our general development about Hilbert spaces suggests a simple question. Can we recover the Hermitian inner product from the squared norm? Some beautiful formulas reveal that the answer is indeed yes. A corresponding result for real vector spaces with inner products does not hold.

To introduce these ideas, let m be an integer with $m \geq 2$. Recall that, for $z \neq 1$, we have

$$1 + z + z^2 + \cdots + z^{m-1} = \frac{1 - z^m}{1 - z}.$$

As we noted in (20) from Chapter I, when z is an m-th root of unity the sum is zero. From Remark I.8.1 the powers of a primitive m-th root of unity may be considered as the elements of a cyclic group Γ of order m.

Let ω be a primitive m-th root of unity and consider averaging the m complex numbers $\gamma \| z + \gamma \zeta \|^2$ as γ varies over Γ. Since each group element is a power of ω, this average equals

$$\frac{1}{m} \sum_{j=0}^{m-1} \omega^j \| z + \omega^j \zeta \|^2.$$

The next proposition gives a simple expression for the average.

Proposition II.4.1. (Polarization identities) Let ω be a primitive m-th root of unity. For $m \geq 3$ we have

$$\langle z, \zeta \rangle = \frac{1}{m} \sum_{j=0}^{m-1} \omega^j \| z + \omega^j \zeta \|^2. \tag{16}$$

For $m = 2$ the right-hand side of (16) equals $2\mathrm{Re}\langle z, \zeta \rangle$.

Proof. The calculations are similar to Exercise 18 from Chapter I and left to the reader. $\qquad\square$

Remark II.4.2. The special case of (16) where $m = 4$ arises often. For $m \geq 3$, each identity expresses the inner product in terms of squared norms. It is both beautiful and useful to recover the inner product from the squared norm.

   We now generalize Proposition II.4.1, and later we apply the result in Chapter IV.

Theorem II.4.3. (Polarization identities for linear transformations) Let $L : \mathcal{H} \to \mathcal{H}$ be a bounded linear transformation. Suppose ω is a primitive m-th root of unity.

1) For $m \geq 3$ we have

$$\langle Lz, \zeta \rangle = \frac{1}{m} \sum_{j=0}^{m-1} \omega^j \langle L(z + \omega^j \zeta), z + \omega^j \zeta \rangle. \tag{17}$$

2) For $m = 2$ the analogous statement is:

$$\langle Lz, \zeta \rangle + \langle L\zeta, z \rangle = \tfrac{1}{2}(\langle L(z+\zeta), z+\zeta \rangle - \langle L(z-\zeta), z-\zeta \rangle). \tag{18}$$

3) Suppose in addition that $\langle Lv, v \rangle$ is real for all $v \in \mathcal{H}$. Then, for all z and ζ,

$$\langle Lz, \zeta \rangle = \overline{\langle L\zeta, z \rangle}.$$

Proof. To prove (17) and (18) expand each $\langle L(z + \omega^j \zeta), z + \omega^j \rangle$ using the linearity of L and of the inner product. Gather all like terms together, and use (20) from Chapter 1. For $m \geq 3$, all terms inside the sum cancel except for m copies of $\langle Lz, \zeta \rangle$. The result gives (17). For $m = 2$, the coefficient of $\langle L\zeta, z \rangle$ does not vanish, and we obtain (18). We have proved statements 1) and 2).

To prove the third statement, we apply the first for some m with $m \geq 3$ and $\omega^m = 1$; the result is

$$\langle Lz, \zeta \rangle = \frac{1}{m} \sum_{j=0}^{m-1} \omega^j \langle L(z + \omega^j \zeta), z + \omega^j \zeta \rangle$$

$$= \frac{1}{m} \sum_{j=0}^{m-1} \omega^j \langle L(\omega^{m-j} z + \zeta), \omega^{m-j} z + \zeta \rangle.$$

Change the index of summation by setting $l = m - j$. Also observe that $\omega^{-1} = \bar{\omega}$. Combining these gives the first equality in (19) below. Finally, because $\langle Lv, v \rangle$ is real, and $\omega^0 = \omega^m$ we obtain the second equality in (19):

$$\langle Lz, \zeta \rangle = \frac{1}{m} \sum_{l=1}^{m} \bar{\omega}^l \langle L(\zeta + \omega^l z), \zeta + \omega^l z \rangle = \overline{\langle L\zeta, z \rangle}. \tag{19}$$

We have now proved the third statement. □

II.5   Generating functions and orthonormal systems

This section offers additional insight into orthonormal systems. It provides valuable information and intuition, but it is logically independent of the rest of the book.

Many of the complete orthonormal systems used in physics and engineering are defined via the *Gram-Schmidt process*. Consider for example an interval I in **R** and the Hilbert space $L^2(I, w(x) \, dx)$ of square integrable functions with respect to some weight function w.

Starting with a nice class of functions, such as the monomials, and then orthonormalizing them, one obtains various *special functions*. The Gram-Schmidt process often leads to tedious computation.

In this section we use the technique of *generating functions* to show how to check whether a collection of functions forms an orthonormal system. We give two standard examples of importance in physics, the Laguerre polynomials and the Hermite polynomials, and introduce more in the exercises.

We begin by proving a simple proposition relating orthonormal systems and generating functions. We then show how the technique works for the Laguerre and Hermite polynomials. The reader should consult [An] for much more information on these polynomials and other special functions. This wonderful book includes at least 21 classes of special functions and 250 identities they satisfy; it also shows how and why they arise in physics and engineering.

Before stating and proving this proposition, it is convenient to extend the notion of convergent power series. Recall that $\mathbf{B}_1$ denotes the open unit disk in $\mathbf{C}$. Suppose $f : \mathbf{B}_1 \to \mathbf{C}$ is a holomorphic function; in Chapter I we remarked that its power series at the origin converges in all of $\mathbf{B}_1$. Let $\mathcal{H}$ be a Hilbert space; it is often useful to consider holomorphic functions on $\mathbf{B}_1$ taking values in $\mathcal{H}$.

Consider a power series $A(z) = \sum A_n z^n$, where the coefficients A_n lie in $\mathcal{H}$. We say that this series converges at the complex number z if its partial sums there form a Cauchy sequence in $\mathcal{H}$. We define a function $A : \mathbf{B}_1 \to \mathcal{H}$ to be holomorphic if there is a sequence $\{A_n\}$ in $\mathcal{H}$ such that the series

$$\sum_{n=0}^{\infty} A_n z^n$$

converges to $A(z)$ for all z in $\mathbf{B}_1$. On compact subsets of $\mathbf{B}_1$ the series converges in norm, and we may therefore rearrange the order of summation at will.

Proposition II.5.1. Let $\mathcal{H}$ be a Hilbert space, and suppose $A : \mathbf{B}_1 \to \mathcal{H}$ is a holomorphic function with $A(t) = \sum_{n=0}^{\infty} A_n t^n$. Then the collection of vectors $\{A_n\}$ forms an orthonormal system in $\mathcal{H}$ if and only if, for all $t \in \mathbf{B}_1$,

$$\| A(t) \|^2 = \frac{1}{1 - |t|^2}.$$

Proof. We have (using the convergence in norm on compact subsets to order the summation as we wish)

$$\| A(t) \|^2 = \sum_{m,n=0}^{\infty} \langle A_n, A_m \rangle t^n \overline{t}^m. \tag{20}$$

Comparison with the geometric series yields the result: the right-hand side of (20) equals $\frac{1}{1-|t|^2}$ if and only if $\langle A_n, A_m \rangle$ equals 0 for $n \neq m$ and equals 1 for $n = m$. $\qquad\qquad\Box$

Suppose that $\{L_n\}$ is a sequence of elements in a Hilbert space. It is sometimes possible to use Proposition II.5.1 to show that $\{L_n\}$ forms an orthonormal system. When $\mathcal{H}$ is a space of functions, and the L_n are known functions, we may be able to find an explicit formula for $A(t)$.

Definition II.5.2. The formal series

$$\sum_{n=0}^{\infty} L_n t^n$$

is the *ordinary generating function* for the sequence $\{L_n\}$. The formal series

$$\sum_{n=0}^{\infty} L_n \frac{t^n}{n!}$$

is the *exponential generating function* for the sequence $\{L_n\}$.

Explicit formulas for these generating functions often provide powerful insight as well as simple proofs of orthogonality relations.

Example II.5.3. (Laguerre polynomials) Suppose that $\mathcal{H} = L^2\big([0, \infty), e^{-x} dx\big)$ is the Hilbert space of square integrable functions on

$[0, \infty)$ with respect to the measure $e^{-x} dx$. Consider functions L_n defined via their generating function by

$$A(x, t) = \sum_{n=0}^{\infty} L_n(x) t^n = (1 - t)^{-1} \exp\left(\frac{-xt}{1 - t}\right). \qquad (21)$$

Note that $x \geq 0$ and $t \in \mathbf{B}_1$. In order to study the inner products $\langle L_n, L_m \rangle$, we compute $|| A(x, t) ||^2$. Formula (21) will enable us to find an explicit formula for the squared norm; Proposition II.5.1 will imply that the L_n form an orthonormal system.

We have

$$| A(x, t) |^2 = (1 - t)^{-1} \exp\left(\frac{-xt}{1 - t}\right) (1 - \bar{t})^{-1} \exp\left(\frac{-x\bar{t}}{1 - \bar{t}}\right).$$

Multiplying by the weight function e^{-x} and integrating we obtain

$$|| A(x, t) ||^2$$

$$= (1 - t)^{-1} (1 - \bar{t})^{-1} \int_0^{\infty} \exp\left(-x\left(1 + \frac{t}{1 - t} + \frac{\bar{t}}{1 - \bar{t}}\right)\right) dx.$$

Computing the integral on the right-hand side and simplifying shows that

$$|| A(x, t) ||^2 = \frac{1}{(1 - t)(1 - \bar{t})} \frac{1}{1 + \frac{t}{1-t} + \frac{\bar{t}}{1-\bar{t}}} = \frac{1}{1 - |t|^2}.$$

From Proposition II.5.1 we see that $\{L_n\}$ forms an orthonormal system in $\mathcal{H}$. □

It is easy to see that the series defining the generating function converges for $|t| < 1$, and it is also evident that each L_n is real-valued. In Exercise 9 we ask the reader to show that the functions L_n satisfy the Rodrigues formula

$$L_n(x) = \frac{e^x}{n!} \left(\frac{d}{dx}\right)^n (x^n e^{-x}) \qquad (22)$$

and hence are polynomials of degree n. They are called the Laguerre polynomials, and they form a *complete* orthonormal system
for $L^2([0, \infty), e^{-x} dx)$. Laguerre polynomials arise in solving the
Schrödinger equation for a hydrogen atom.

A similar technique works for the Hermite polynomials, which
arise in many problems in physics, such as the simple harmonic oscil-
lator. One way to define the Hermite polynomials is via the exponential
generating function

$$\exp(2xt - t^2) = \sum H_n(x)\frac{t^n}{n!}. \tag{23}$$

The functions H_n are polynomials and form an orthogonal set for
$\mathcal{H} = L^2(\mathbf{R}, e^{-x^2} dx)$. With this normalization the norms are not equal
to unity. In Exercise 11 the reader is asked to study the Hermite polynomials by mimicking the computations for the Laguerre polynomials.
The reader should be aware that other normalizations of these polynomials are also common. For example sometimes the weight function
used is $e^{\frac{-x^2}{2}}$.

Remark II.5.4. When the Hilbert space consists of square-integrable
functions on a domain in $\mathbf{C}^n$, it is sometimes useful to replace the single complex variable $t \in \mathbf{C}$ by a variable $t \in \mathbf{C}^n$. Then one indexes the
countably infinite collection of vectors by the multi-indices in n variables. There is a simple analogue of Proposition II.5.1, whose statement
and proof we leave to the reader in Exercise 10 of Chapter III.

Exercise 9.

1) With L_n as in Example II.5.3, verify the Rodrigues formula (22).
   Suggestion: Write the power series of the exponential on the right-
   hand side of (21) and interchange the order of summation.

2) Show that each L_n is a polynomial in x. Hint: The easiest way is to
   use 1).

3) Prove that $\{L_n\}$ forms a *complete* system in $L^2\left([0, \infty), e^{-x}\, dx\right)$.

Exercise 10. For $x > 0$ verify that

$$\sum_{n=0}^{\infty} \frac{L_n(x)}{n+1} = \int_0^{\infty} \frac{e^{-xt}}{t+1}\, dt.$$

Suggestion: Integrate the relation

$$\sum_{n=0}^{\infty} L_n(x) s^n = (1-s)^{-1} \exp\left(\frac{-xs}{1-s}\right)$$

over the interval $[0, 1]$ and then change variables in the integral.

Exercise 11. (**Hermite polynomials**) Here H_n is defined by (23).

1) Use (23) to find a simple expression for

$$\sum_{n=0}^{\infty} H_n(x) \frac{t^n}{n!} \sum_{m=0}^{\infty} H_m(x) \frac{s^m}{m!}.$$

2) Integrate the result in 1) over $\mathbf{R}$ with respect to the measure $e^{-x^2}\, dx$.

3) Use 2) to show that the Hermite polynomials form an orthogonal system with

$$\| H_n \|^2 = 2^n n! \sqrt{\pi}.$$

4) Prove that the system of Hermite polynomials is complete in $L^2(\mathbf{R}, e^{-x^2}\, dx)$.

Exercise 12. Define the Legendre polynomials $P_n(x)$ for $|x| \le 1$ and $|t| < 1$ by

$$\frac{1}{\sqrt{1 - 2xt + t^2}} = \sum_{n=0}^{\infty} P_n(x) t^n.$$

Prove that P_n is a polynomial of degree n and that the collection of these polynomials forms a complete orthogonal system for $L^2([-1, 1], dx)$. Compute $\| P_n \|^2$.

Exercise 13. Replace the generating function in Exercise 12 by $(1 - 2xt + t^2)^{-\lambda}$ for $\lambda > -\frac{1}{2}$ and carry out the same steps. The resulting polynomials are the *ultraspherical* or *Gegenbauer* polynomials.

CHAPTER III

Complex Analysis in Several Variables

The material in this chapter intentionally interrupts our theoretical development about Hilbert spaces. There are at least two reasons for this interruption. One reason is that this material is a prerequisite for understanding the Hilbert space $A^2(\mathbf{B}_n)$ of square-integrable holomorphic functions on the unit ball; this particular Hilbert space arises in a crucial way in the proof of Theorem VII.1.1, the main result in the book. A second reason is the feeling that abstract material doesn't firmly plant itself in one's mind unless it is augmented by concrete material. Conversely the presentation of concrete material benefits from appropriate abstract interludes.

In this chapter we introduce holomorphic functions of several complex variables. This presentation provides only a brief introduction to the subject. Multi-index notation and issues involving calculus of several variables also appear here. Studying them allows us to provide a nice treatment of the gamma and beta functions. We use them to compute the Bergman kernel function for the unit ball, thereby reestablishing contact with Hilbert spaces.

III.1   Holomorphic functions

Our study of $A^2(\mathbf{B}_n)$ requires us to first develop some basic information about holomorphic functions of several complex variables. As in one variable, holomorphic functions of several variables are locally represented by convergent power series. Although some formal aspects of the theories are the same, geometric considerations change considerably in the higher-dimensional theory.

In order to avoid a profusion of indices we first introduce multi-index notation. This notation makes many computations in several variables both easier to perform and simpler to expose.

Suppose $z = (z_1, \ldots, z_n) \in \mathbf{C}^n$, and let $(\alpha_1, \ldots, \alpha_n)$ be an n-tuple of nonnegative integers. We call α a *multi-index* and write

$$z^\alpha = \prod_{j=1}^{n} z_j^{\alpha_j}. \tag{1}$$

Notice that the monomial defined in (1) is of degree $\sum \alpha_j$. So it is natural to write

$$|\alpha| = \sum \alpha_j$$

for the *order* or *length* of the multi-index α. Already we obtain nice notation for polynomials. Let q be a polynomial of degree d in n complex variables. Then, where each coefficient c_α is a complex number, and where $c_\alpha \neq 0$ for some α with $|\alpha| = d$,

$$q(z) = \sum_{|\alpha|=0}^{d} c_\alpha z^\alpha.$$

A polynomial q on $\mathbf{C}^n$ is *homogeneous* of degree m if

$$q(tz) = t^m q(z)$$

for all $t \in \mathbf{C}$ and for all $z \in \mathbf{C}^n$. Observe that a polynomial q is homogeneous of degree m if and only if

$$q(z) = \sum_{|\alpha|=m} c_\alpha z^\alpha.$$

Convergent power series are limits of polynomials. In several variables the notion of partial sum is more elusive than it is in one variable. Given a power series, we let $q_m(z)$ denote the terms in the series that are homogeneous of degree m. Then $\sum_{m=0}^{N} q_m(z)$ plays the role of the N-th partial sum:

$$\sum_{|\alpha|=0}^{N} c_\alpha z^\alpha = \sum_{m=0}^{N} \sum_{|\alpha|=m} c_\alpha z^\alpha = \sum_{m=0}^{N} q_m(z).$$

We say that the power series $\sum_\alpha c_\alpha z^\alpha$ converges at z if the limit, as N tends to infinity, of this N-th partial sum exists at z. A convergent power series therefore has an expansion in terms of homogeneous parts:

$$\sum_{|\alpha|=0}^{\infty} c_\alpha z^\alpha = \sum_{m=0}^{\infty} q_m(z) = \lim_{N \to \infty} \sum_{m=0}^{N} q_m(z). \qquad (2)$$

The proofs of the following important statements are left as an exercise. Suppose that the power series (2) converges at z. Then it converges for all w such that $|w_j| < |z_j|$ for all j. Suppose that the power series (2) converges at each point of an open set Ω. Then it converges uniformly and absolutely on any compact subset of Ω. As a consequence, (2) determines an infinitely differentiable function on Ω.

Exercise 1. Prove the statements made in the previous paragraph about convergence of power series.

The meaning of many mathematical terms depends on context; the word *domain* provides an example. The *domain* of a function is the set where it is defined. A *domain* in a metric space is an open connected subset. These two uses of the word merge when we talk about holomorphic functions; such functions are defined on connected open sets in the metric space $\mathbf{C}^n$.

Definition III.1.1. Let Ω be a domain in $\mathbf{C}^n$, and suppose $f : \Omega \to \mathbf{C}$ is a function. We say that f is *holomorphic* on Ω if, for each p in Ω, there is an open ball $B_r(p)$ in Ω and a power series based at p,

$$\sum_{|\alpha|=0}^{\infty} c_\alpha (z - p)^\alpha, \qquad (3)$$

that converges to $f(z)$ for all $z \in B_r(p)$.

Example III.1.2. Any polynomial in the variables $(z_1, \ldots, z_n)$ is a holomorphic function on $\mathbf{C}^n$. The quotient of two polynomials in z is holomorphic on the complement of the zero-set of the denominator. Let f be a holomorphic function on $\mathbf{C}^n \times \mathbf{C}^n$. Then the function $z \rightarrow f(z, \bar{z})$ is *not* a holomorphic function unless it is independent of the $\bar{z}$ variable. See Remark III.1.3. This type of function does, however, play a major role in this book.

Let f be a holomorphic function of two complex variables z_1 and z_2. For z_2 fixed, the function $z_1 \rightarrow f(z_1, z_2)$ is a holomorphic function of one variable. There is a famous converse assertion due to Hartogs. Suppose that f is a (not necessarily continuous) function of two complex variables, and further suppose that f is holomorphic in each variable separately (when the other variable is fixed). Then f is holomorphic. See [H] for a proof. By contrast, consider the function f of two real variables given by $f(0, 0) = 0$ and by

$$f(x, y) = \frac{xy}{x^2 + y^2}$$

away from $(0, 0)$. For arbitrary fixed x or y, this function is infinitely differentiable in the other variable. As a function of two variables it is not even continuous at the origin!

Remark III.1.3. Holomorphic functions are locally represented by convergent power series. In particular holomorphic functions in several variables depend on the z variables, but not on the $\bar{z}$ variables. Thus holomorphic functions are solutions to the system of partial differential equations $\frac{\partial f}{\partial \bar{z}_j} = 0$ for $j = 1, \ldots, n$. This viewpoint has played a dominant role in complex analysis in several variables since 1960. See [H] and [Kr]; see also [GK] for a treatment of complex analysis in one variable from this point of view.

An interesting aspect of function theory in several variables is that balls are no longer the only domains of convergence for a power series. This issue will not matter in this book, but we mention here that the region of convergence of a power series in several variables can be much more complicated than a ball. We illustrate this situation with some examples and with a general result that is stated without proof.

Example III.1.4. (Regions of convergence)

2.1). Consider $\sum_{k=0}^{\infty} z_1^k z_2^k$. This series is a geometric series in the product $z_1 z_2$, and hence the region of convergence is $\{(z_1, z_2) : |z_1 z_2| < 1\}$.

2.2). Suppose that $f(z_1, z_2) = g(z_1)h(z_2)$, where g and h are power series in one complex variable. Then the region of convergence for f is the Cartesian product of the regions of convergence for the factors.

2.3). See [H] for a proof of the following result. Let D^* be an open convex set in $\mathbf{R}^n$. Suppose that whenever $t \in D^*$, $s \in \mathbf{R}^n$, and $s_j \leq t_j$ for all j, then $s \in D^*$. Define an open domain D in $\mathbf{C}^n$ by $z \in D$ if and only if $|z_j| \leq e^{t_j}$ for all j and some $t \in D^*$. Then D is the domain of convergence of some power series.

On occasion we will be interested in holomorphic mappings. These are sometimes called *vector-valued* holomorphic functions. In Proposition II.5.1 we encountered holomorphic functions from $\mathbf{B}_1$ to a Hilbert space. Let now Ω be a domain in $\mathbf{C}^n$, let $\mathcal{H}$ be a Hilbert space, and let $f : \Omega \to \mathcal{H}$ be a function. As before we say that f is a holomorphic mapping on Ω if, for each $p \in \Omega$, it can be locally represented by a convergent power series based at p in an open set containing p. Convergence means that the sequence $\{S_N(z)\}$, defined by

$$S_N(z) = \sum_{|\alpha|=0}^{N} c_\alpha (z - p)^\alpha,$$

is a Cauchy sequence in $\mathcal{H}$. In case the Hilbert space is complex Euclidean space, then f is a holomorphic mapping if and only if its components are holomorphic (complex-valued) functions. Later in this book we will often consider holomorphic mappings $f : \mathbf{C}^n \to \mathbf{C}^k$ and then study their squared norms $\| f \|^2$.

We return to complex-valued holomorphic functions. Let Ω be a domain in $\mathbf{C}^n$. We write $\mathcal{O}(\Omega)$ for the collection of holomorphic functions $f : \Omega \to \mathbf{C}$. Then $\mathcal{O}(\Omega)$ is a ring; the sum and product of holomorphic functions are also holomorphic. Furthermore, all derivatives of a holomorphic function exist and define holomorphic functions. The ring $\mathcal{O}(\Omega)$ has other nice properties, to which we now turn.

See [Kr] or [H] for a proof of the following result.

Theorem III.1.5. Suppose that $\{f_n\}$ is a sequence in $\mathcal{O}(\Omega)$, and that, for each compact subset $K \subset \Omega$, the sequence $\{f_n\}$ converges uniformly on K. Then the limit function f lies in $\mathcal{O}(\Omega)$.

Holomorphic functions are closely related to harmonic functions. Let Ω be a domain in $\mathbf{R}^m$, and suppose $u : \Omega \to \mathbf{R}$ is a twice differentiable function. Then u is called *harmonic* if it is a solution of the Laplace equation

$$\sum_{k=1}^{m} \frac{\partial^2 u}{\partial x_k^2} = 0. \tag{4}$$

Functions satisfying (4) are automatically infinitely differentiable. See for example [Ta]. An alternative characterization is that a continuous function is harmonic if and only if it satisfies the *mean-value property* (5):

$$u(x) = \frac{1}{V(B_r(x))} \int_{B_r(x)} u(t) \, dV(t)$$

$$= \frac{1}{c_m r^m} \int_{B_r(0)} u(x + y) \, dV(y). \tag{5}$$

Thus, for each point x, and each open ball about x lying in the domain of u, the value $u(x)$ at the center of the ball equals the average value of u over that ball. In the first integral $dV(t)$ denotes the usual m-dimensional volume element, and $V(B_r(x))$ denotes the volume of the ball $B_r(x)$. In the next section we will show how to find the constant c_m in the formula $c_m r^m$ for the volume of a ball of radius r in

m-dimensional space. A continuous function satisfying (5) is necessarily infinitely differentiable and satisfies (4).

We briefly discuss the relationship between holomorphic and harmonic functions. Suppose that $m = 2n$, and we put $z_j = x_j + ix_{n+j}$ for $j = 1, \ldots, n$. This identification enables us to view $\mathbf{R}^{2n}$ as $\mathbf{C}^n$. The Laplace operator can be expressed in terms of z and $\bar{z}$:

$$\sum_{k=1}^{2n} \frac{\partial^2}{\partial x_k^2} = 4 \sum_{j=1}^{n} \frac{\partial^2}{\partial z_j \partial \bar{z}_j}. \tag{6}$$

It follows easily from Remark III.1.3 and (6) that the real and imaginary parts of a holomorphic function f (in any number of variables) are harmonic functions. In one variable, a harmonic function on a ball, for example, is necessarily the real part of a holomorphic function. In more than one variable no such converse assertion is true.

Example III.1.6. Put $u(z) = |z_1|^2 - |z_2|^2$. Then u is harmonic on $\mathbf{C}^2$, but it is not the real or imaginary part of a holomorphic function.

Holomorphic functions satisfy a mean-value property as well. For each z in the domain of f, and each ball $B_r(z)$ contained in the domain of f, we have

$$f(z) = \frac{1}{V(B_r(z))} \int_{B_r(z)} f(w) \, dV(w).$$

This conclusion follows immediately from the mean-value properties of the harmonic functions $\operatorname{Re}(f)$ and $\operatorname{Im}(f)$. The mean-value property for holomorphic functions arises in the proofs of many results in complex analysis.

Our next result is crucial in the development of the book. Using the mean-value property we show that estimates for holomorphic functions in $L^2(\Omega)$ imply estimates in the supremum norm. Two corollaries are used many times in the sequel. Corollary III.1.8 guarantees that $A^2(\Omega)$ is a Hilbert space, and Corollary III.1.9 shows that evaluation at a point is a continuous linear functional on $A^2(\Omega)$. These facts allow us to apply the Riesz representation lemma (Theorem II.2.4) and thereby define the Bergman kernel function in Section 3.

Proposition III.1.7. Let Ω be a bounded domain in $\mathbf{C}^n$, and let K be a compact subset of Ω. Then there is a constant C_K such that, whenever $f \in A^2(\Omega)$,

$$\sup_{z \in K} |f(z)| \leq C_K \|f\|. \tag{7}$$

Here $\|f\|$ denotes the L^2 norm of f.

Proof. For each $z \in K$, the minimum distance from z to the boundary of Ω is a positive number. Because K is compact, the collection of such positive numbers is bounded away from zero. So we may choose $r > 0$ such that, for each $z \in K$, the Euclidean ball $B_r(z)$ lies in Ω. The mean-value property for f on that ball gives

$$f(z) = \frac{1}{V(B_r(z))} \int_{B_r(z)} f(w)\, dV(w).$$

We estimate the integral using the Cauchy-Schwarz inequality, applied to f and 1. Since the volume $V = V(B_r(z))$ is the same for all the balls, we obtain

$$\sup_{z \in K} |f(z)| \leq \frac{1}{V} \sup_z \int_{B_r(z)} |f(w)|\, dV(w)$$

$$\leq \frac{1}{V} \left(\int_{B_r(z)} |f|^2\, dV \right)^{\frac{1}{2}} V^{\frac{1}{2}} \leq \frac{1}{\sqrt{V}} \|f\|.$$

This proves the desired inequality. $\qquad\square$

Corollary III.1.8. $A^2(\Omega)$ is a closed subspace of $L^2(\Omega)$ and hence a Hilbert space.

Proof. It is clear that $A^2(\Omega)$ is a subspace. The issue is to show that it is closed. To do so, let $\{f_\nu\}$ be a Cauchy sequence of holomorphic

functions in $A^2(\Omega)$. Then $\{f_\nu\}$ converges to a limit in $L^2(\Omega)$. We apply
Proposition III.1.7 to $f_\nu - f_\mu$ and conclude that $\{f_\nu\}$ converges uniformly on compact subsets of Ω. By Theorem III.1.5 the limit function
is also holomorphic, and hence in $A^2(\Omega)$. Therefore $A^2(\Omega)$ is closed.

$\square$

Corollary III.1.9. Choose $z \in \Omega$. Define a mapping $L : A^2(\Omega) \to \mathbf{C}$
by $L(f) = f(z)$. Then L is a bounded linear functional on $A^2(\Omega)$.

Proof. The linearity of L is clear. The boundedness is easy. Since the
single point z is a compact subset of Ω, (7) yields

$$|L(f)| = |f(z)| \le C_z \|f\|$$

and hence L is bounded.

$\square$

The next exercise, which only gives a glimpse of the importance
of the Poisson kernel, provides a nice application of the mean-value
property for harmonic functions.

Exercise 2. The Poisson kernel for the unit disk in $\mathbf{C}$.

1) For $w \ne 1$ verify the identity

$$\mathrm{Re}\left(\frac{1+w}{1-w}\right) = \frac{1-|w|^2}{|1-w|^2}.$$

2) Use the mean-value property for harmonic functions to show that,
for $0 \le R < 1$,

$$\pi R^2 = \int_{|w| \le R} \frac{1-|w|^2}{|1-w|^2}\, dV(w). \tag{8}$$

3) Write the right-hand side of (8) as an integral in polar coordinates,
and then use the fundamental theorem of calculus to show that, for

$0 \leq R < 1$,

$$2\pi = \int_0^{2\pi} \frac{1 - R^2}{|1 - Re^{i\theta}|^2} d\theta.$$

Exercise 3. Suppose that u is a function of two real variables and that there is a holomorphic function f of one complex variable such that $u(x, y) = \operatorname{Re} f(z)$. Thus u is harmonic, and therefore u is locally represented by a convergent power series in the real variables x and y. Prove (9):

$$f(z) = 2u\left(\frac{z}{2}, \frac{z}{2i}\right) - \overline{f(0)}. \tag{9}$$

Formula (9) shows that we can recover f, up to an additive constant, from u, by substituting certain complex variables for the real variables in the formula for u. In standard complex variables courses one finds f via a process involving both differentiation and integration. The point of the exercise is that one can find f directly by substitution! Check (9) when $u(x, y) = x^2 - y^2$ and when $u(x, y) = e^x \cos(y)$.

Exercise 4. What are the solutions u to the system of partial differential equations given by $u_{z_i \bar{z}_j} = 0$ for all i, j?

III.2   Some calculus

Recall that $\mathbf{B}_n$ denotes the open unit ball in $\mathbf{C}^n$. We will show that the collection of normalized monomials is a complete orthonormal system for $A^2(\mathbf{B}_n)$. This example both illustrates the notion of complete orthonormal system and plays a fundamental but subtle role in the subsequent development.

We first consider some notations and facts from calculus of several variables. It is crucial to enhance our skill with multi-index notation. Recall that z^α is defined by (1); it is also useful to write $(zw)^\alpha$ instead of $z^\alpha w^\alpha$. When α is a multi-index we write $\alpha!$ for $\prod(\alpha_j!)$. It is convenient to employ these conventions when α is an n-tuple of nonnegative numbers. Our work on the gamma function in this section will make precise this extended sense of factorials.

An expression such as $z^{2\alpha-1}$ has meaning; here $2\alpha - 1$ is the multi-index whose j-th entry is $2\alpha_j - 1$. In this case 1 denotes a multi-index consisting of n ones; according to our conventions the strange equality $|1| = n$ results! The context prevents confusion.

Multi-index notation also applies to differentials. For example, it is natural to write $dx = dx_1 dx_2 \ldots dx_n = dV(x)$ for the volume form associated with Lebesgue measure on $\mathbf{R}^n$. For another example, we consider the volume form in polar coordinates in $\mathbf{C}^n$. Thus we make the usual polar coordinate change $z_j = r_j e^{i\theta_j}$ in each variable. Using multi-index notation we write instead $z = re^{i\theta}$, where $r \in \mathbf{R}^n$ and each $r_j \geq 0$, and $\theta \in [0, 2\pi)^n$. The volume form $dV(z)$ for Lebesgue measure becomes $r \, dr \, d\theta$. In other words, $dV(z) = r \, dr \, d\theta = \prod_{j=1}^n r_j \, dr_j \, d\theta_j$.

We next define the gamma function, which enables us to meaningfully write $x!$ when x is a real number with $x > -1$. This special function arises in probability and statistics, as well as in several crucial computations in this book.

Definition III.2.1. (**gamma function**) Suppose $x > 0$. We put

$$\Gamma(x) = \int_0^\infty e^{-t} t^{x-1} \, dt.$$

Observe that the integral defining $\Gamma(x)$ is *improper*, because infinity is a limit of integration. Observe further that the integral is also improper at zero when $0 < x < 1$, because the integrand is infinite there. Nevertheless the integral converges.

The gamma function generalizes the factorial function; we see this using integration by parts. When k is a positive integer, $\Gamma(k) = (k-1)!$. When x is a positive real number, $\Gamma(x + 1) = x\Gamma(x)$. The gamma function helps us to evaluate an innocent-looking definite integral.

Lemma III.2.2. (**Euler beta function**) Suppose that a and b are positive numbers. The following formula holds:

$$\int_0^1 u^{a-1}(1-u)^{b-1}\, du = \frac{\Gamma(a)\Gamma(b)}{\Gamma(a+b)}.$$

Proof. We give only a sketch. The reader should justify each statement. We have

$$\Gamma(a)\Gamma(b) = \int_0^\infty \int_0^\infty e^{-(s+t)} s^{a-1} t^{b-1}\, ds\, dt$$

$$= \int_0^\infty \int_0^\infty e^{-(S^2+T^2)} 4 S^{2a-1} T^{2b-1}\, dS\, dT$$

after making the change of variables $(s, t) = (S^2, T^2)$. Then use polar coordinates by putting $(S, T) = (r\cos(\theta), r\sin(\theta))$. This change of variables yields

$$\Gamma(a)\Gamma(b) = \int_0^\infty \int_0^{\frac{\pi}{2}} e^{-r^2} r^{2(a+b)-1} 4 (\cos(\theta))^{2a-1} (\sin(\theta))^{2b-1}\, d\theta\, dr.$$

$$(10)$$

The double integral in (10) separates into a product of two integrals. Group a factor of 2 with each. Then the first integral is easily seen to be $\Gamma(a+b)$. The second integral is

$$2 \int_0^{\frac{\pi}{2}} \cos^{2a-1}(\theta) \sin^{2b-1}(\theta)\, d\theta.$$

Change variables by putting $u = \cos^2(\theta)$. This yields the integral $\int_0^1 u^{a-1}(1-u)^{b-1}\, du$. Dividing by $\Gamma(a+b)$ gives the result.   □

The function of the two variables a and b from Lemma III.2.2 is called the Euler beta function. It often arises in probability and statistics. Its generalization to more variables will help us evaluate certain norms defined by integrals.

Definition III.2.3. Let α be an n-tuple of positive numbers, and let $|\,\alpha\,|$ denote their sum. Let K denote the subset of $\mathbf{R}^n$ defined by $r_j \geq 0$ for all j and $\sum r_j^2 \le 1$. We put

$$\mathcal{B}(\alpha) = 2^n |\alpha| \int_K r^{2\alpha-1} \, dr.$$

When $n = 1$, we have $\mathcal{B}(a) = 1$ for all a. When $n = 2$ we write $\alpha = (a, b)$. Computing the inner integral of the double integral (and leaving a few steps for the reader) yields

$$\mathcal{B}(a, b) = 4(a + b) \int_0^1 \int_0^{\sqrt{1-x^2}} y^{2b-1} x^{2a-1} dy \, dx$$

$$= \frac{a+b}{b} \int_0^1 u^{a-1}(1 - u)^b \, du = \frac{\Gamma(a)\Gamma(b)}{\Gamma(a+b)}.$$

Therefore $\mathcal{B}(a, b)$ gives the Euler beta function. Definition III.2.3 of the beta function of n variables exhibits this function as an n-dimensional integral. When $n = 2$ we obtain a single integral after integrating the radial variable.

As an analogue to our convention for $\alpha!$, it is natural to write $\Gamma(\alpha)$ for the product $\prod \Gamma(\alpha_j)$. This typographical convention helps lead to a beautiful formula generalizing Lemma III.2.2.

Lemma III.2.4. Let α be an n-tuple of positive numbers. Then

$$\mathcal{B}(\alpha) = \frac{\Gamma(\alpha)}{\Gamma(|\alpha|)} = \frac{\prod \Gamma(\alpha_j)}{\Gamma(\sum \alpha_j)}. \tag{11}$$

Proof. Start by writing $\prod \Gamma(\alpha_j)$ as an iterated integral. Imitate the proof of Lemma III.2.2 by making appropriate changes of variables. The details are left to the reader. □

This formula enables us to compute the squared norm in $A^2(\mathbf{B}_n)$ of each monomial. We have the following fundamental fact:

Proposition III.2.5. The normalized monomials $\frac{z^\alpha}{c_\alpha}$ define a complete orthonormal system in $A^2(\mathbf{B}_n)$. Here

$$c_\alpha^2 = \frac{\pi^n \alpha!}{(n + |\alpha|)!} = \frac{\pi^n \prod \alpha_j!}{(n + \sum \alpha_j)!}.$$

Proof. integrating using polar coordinates in each variable separately. Thus $z = re^{i\theta}$ in multi-index notation. Let T denote $[0, 2\pi]^n$; let K denote the set of r in $\mathbf{R}^n$ such that $r_j \geq 0$ for each j and $\sum r_j^2 \leq 1$. The change of variables formula for multiple integrals yields:

$$\langle z^\alpha, z^\beta \rangle = \int_{\mathbf{B}_n} z^\alpha \bar{z}^\beta \, dV(z) = \int_T \int_K e^{i(\alpha - \beta)\theta} r^{\alpha + \beta + 1} \, dr \, d\theta. \quad (12)$$

When $\alpha \neq \beta$, the value of (12) is 0, because of the θ integration. When $\alpha = \beta$, the θ integration gives $(2\pi)^n$. We obtain

$$\langle z^\alpha, z^\alpha \rangle = (2\pi)^n \int_K r^{2\alpha + 1} \, dr = \frac{\pi^n B(\alpha + 1)}{|\alpha + 1|} = \frac{\pi^n \alpha!}{|\alpha + 1| \Gamma(|\alpha + 1|)}$$

$$= \frac{\pi^n \alpha!}{(n + |\alpha|)!} = c_\alpha^2$$

by using the generalized beta function and Lemma III.2.4.

It remains to verify the completeness of the set of monomials. The strategy is to choose an $f \in A^2(\Omega)$ that is orthogonal to all monomials, and then to prove that $f = 0$. We assume without proof one fact from complex analysis: a holomorphic function on $\mathbf{B}_n$ is given by a power series in z that converges uniformly on compact subsets of $\mathbf{B}_n$.

Choose ϵ with $0 < \epsilon < 1$. Let I_ϵ denote $\{z : ||z|| \leq 1 - \epsilon\}$, and let J_ϵ denote $\{z : 1 - \epsilon < ||z|| < 1\}$. Note that the volume of J_ϵ tends to zero as ϵ tends to zero.

Suppose $f \in A^2(\mathbf{B}_n)$ and $\langle f, z^\alpha \rangle = 0$ for all α. For each α we have

$$0 = \langle f, z^\alpha \rangle = \int_{I_\epsilon} f(z) \bar{z}^\alpha \, dV(z) + \int_{J_\epsilon} f(z) \bar{z}^\alpha \, dV(z).$$

The integrals over the two pieces therefore have the same absolute value. Using this, the estimate $|z^\alpha| \leq 1$, and the Cauchy-Schwarz inequality gives

$$\left| \int_{I_\epsilon} f(z) \bar{z}^\alpha \, dV(z) \right| = \left| \int_{J_\epsilon} f(z) \bar{z}^\alpha \, dV(z) \right| \leq c(\epsilon) \, \| f \|_2,$$

where $c(\epsilon)$ is the square root of $\text{Vol}(J_\epsilon)$. Hence $c(\epsilon)$ tends to zero as ϵ tends to zero.

Since $f(z) = \lim_{N \to \infty} \sum_{|\mu| \leq N} f_\mu z^\mu$, and the convergence is uniform on I_ϵ, we may interchange the limit and the integral over I_ϵ. Distinct monomials are orthogonal on I_ϵ, so the only contribution arises when $\mu = \alpha$. Hence

$$\left| f_\alpha \int_{I_\epsilon} |z^\alpha|^2 \, dV(z) \right| \leq c(\epsilon) \, \| f \|_2.$$

Letting ϵ tend to zero shows that $| f_\alpha c_\alpha^2 | \leq 0$, and hence $f_\alpha = 0$. Since α is an arbitrary multi-index, all coefficients in the power series for f vanish, so $f = 0$. By Definition II.3.1 the collection of normalized monomials forms a complete system. $\qquad \square$

Let W_α denote the one-dimensional subspace of $A^2(\mathbf{B}_n)$ generated by z^α. Then $A^2(\mathbf{B}_n)$ is the orthogonal sum of the W_α. It is convenient to consider, for $m \geq 0$, the subspace V_m of homogeneous polynomials of degree m. The dimension of V_m is the binomial coefficient $\binom{n+m-1}{m}$. It will be also useful to express $A^2(\mathbf{B}_n)$ as the orthogonal sum of the V_m.

Exercise 5. Use the integral from Definition III.2.1 to define $\Gamma(z)$ for a complex number z with positive real part. Verify that $\Gamma(z+1) = z\Gamma(z)$, and use this formula to extend the definition of Γ to as large a set as possible. Show that this extended version of the gamma function is holomorphic in $\mathbf{C}$ except at the nonpositive integers.

Exercise 6. Fill in the details of the computation in Lemma III.2.2. Then prove Lemma III.2.4 by a similar method.

Exercise 7. Use Definition III.2.3 and Lemma III.2.4 to find a formula for the volume of the unit ball in n-dimensional real Euclidean space. Hint: Put $\alpha = (\frac{1}{2}, \ldots, \frac{1}{2})$.

III.3   The Bergman kernel function

The notion of orthogonal projection leads to the Bergman kernel function. Let Ω be a bounded domain in $\mathbf{C}^n$. Recall from Corollary III.1.8 that $A^2(\Omega)$ is a closed subspace of $\mathcal{H} = L^2(\Omega)$. The Bergman kernel function provides an integral formula for the orthogonal projection onto $A^2(\Omega)$ defined in Corollary II.2.3.

Fix $z \in \Omega$ and define a linear functional K_z on $A^2(\Omega)$ by $K_z f = f(z)$. The linear functional K_z is bounded on $A^2(\Omega)$ by Corollary III.1.9. By the Riesz representation lemma (Theorem II.2.4) there is a unique $\delta_z \in A^2(\Omega)$ such that $K_z f = \langle f, \delta_z \rangle$ for all $f \in A^2(\Omega)$.

Let Ω^* denote the complex conjugate domain of Ω; thus Ω^* consists of those $\overline{w} \in \mathbf{C}^n$ such that $w \in \Omega$. Let r be a function on $\Omega \times \Omega^*$. We denote the value of r at the point $(z, \overline{w})$ by $r(z, \overline{w})$; this notation emphasizes the dependence on $\overline{w}$, and will be especially helpful in the sequel. The function δ_z is defined and holomorphic on Ω. We may view its complex conjugate as defined on Ω^*. Letting z vary we define a function on $\Omega \times \Omega^*$ by

$$B(z, \overline{w}) = \overline{\delta_z(w)}. \tag{13}$$

Definition III.3.1. The Bergman kernel function of a bounded domain Ω in $\mathbf{C}^n$ is the function $B : \Omega \times \Omega^* \to \mathbf{C}$ defined by (13).

Lemma III.3.2. The Bergman kernel function satisfies the following conjugate symmetry property: $B(z, \overline{w}) = \overline{B(w, \overline{z})}$.

Proof. This property follows formally and easily from (13):

$$\overline{B(z, \overline{w})} = \delta_z(w) = \langle \delta_z, \delta_w \rangle = \overline{\langle \delta_w, \delta_z \rangle} = B(w, \overline{z}).$$

The result follows by taking conjugates. $\square$

For z fixed, δ_z is in $A^2(\Omega)$ as a function of w. Therefore, again with z fixed, the Bergman kernel is the complex conjugate of a square-integrable holomorphic function in w. By Lemma III.3.2, the Bergman kernel function is therefore in $A^2(\Omega)$ as a function of z, for fixed w. Two additional facts about the Bergman kernel function also appear in our list of four key properties:

B1) $B(z, \overline{w}) = \overline{B(w, \overline{z})}$.

B2) For w fixed, $B(z, \overline{w})$ is holomorphic as a function of z and in $A^2(\Omega)$.

B3) For all $h \in A^2(\mathbf{B}_n)$, we have

$$h(z) = \langle h, \delta_z \rangle = \int_{\Omega} h(w) B(z, \overline{w}) \, dV(w).$$

B4) Suppose $f \in L^2(\Omega)$, and f is orthogonal to $A^2(\Omega)$. Then

$$0 = \langle f, \delta_z \rangle = \int_{\Omega} f(w) B(z, \overline{w}) \, dV(w).$$

We have verified properties B1) and B2) already. Properties B3) and B4) are immediate from (13) and the properties of δ_z.

When z is fixed, $B(z, \overline{w})$ is the complex conjugate of a holomorphic function of w. Such functions are known as *anti-holomorphic* or *conjugate-holomorphic* functions; they are locally given near a point p by convergent power series in $\overline{w - p}$.

Property B3) is known as the *reproducing property* of the Bergman kernel. It provides an example of an integral representation formula. One should not get too excited here; B3) expresses the value of f at z in terms of the values of f on all of Ω. Furthermore, for most domains the Bergman kernel is not known explicitly. Theorem III.3.5 gives an explicit formula for the Bergman kernel function of the unit ball.

The next result shows that properties B1), B2), and B3) uniquely determine the Bergman kernel function. In particular any function satisfying these three properties also satisfies Property B4).

Lemma III.3.3. There is a unique function satisfying properties B1), B2), and B3).

Proof. Suppose that both $B(z, \overline{w})$ and $C(z, \overline{w})$ satisfy properties B1), B2), and B3). The following string of equalities then holds:

$$
\begin{aligned}
B(z, \overline{w}) &= \int_{\Omega} B(\zeta, \overline{w}) C(z, \overline{\zeta}) \, dV(\zeta) \\
&= \overline{\int_{\Omega} B(w, \overline{\zeta}) C(\zeta, \overline{z}) \, dV(\zeta)} = \overline{C(w, \overline{z})} = C(z, \overline{w}).
\end{aligned}
$$

The first equality uses property B3) for C on the function B, holomorphic in its first slot by B2). The second equality uses property B1) for both B and C. The third equality uses property B3) for B on the function C, holomorphic in its first slot by B2). $\square$

There is an intriguing formula connecting the Bergman kernel function with the notion of complete orthonormal system. We will use it to compute the Bergman kernel function for the unit ball.

Proposition III.3.4. Let Ω be a bounded domain in $\mathbf{C}^n$, and suppose that $\{\phi_\alpha\}$ is a complete orthonormal system for $A^2(\Omega)$. Consider the sum

$$
\sum_{\alpha} \phi_\alpha(z) \overline{\phi_\alpha(w)}. \tag{14}
$$

This sum converges uniformly on compact subsets of $\Omega \times \Omega^*$ to the Bergman kernel function $B(z, \overline{w})$, and is therefore independent of the orthonormal system used.

Proof. The summation index used in (14) is denoted α because our application of this result has an orthonormal system indexed by the multi-indices. In this proof, however, it is convenient to replace the multi-index α by the positive integer j.

Fix w and consider the sum of terms in $A^2(\Omega)$ defined by

$$\sum_{j=1}^{\infty} \phi_j \overline{\phi_j(w)}. \tag{15}$$

We will show first that the partial sums of (15) form a Cauchy sequence in $A^2(\Omega)$. Choose positive integers N and M with $N < M$. Orthonormality implies

$$\left\| \sum_{j=N+1}^{M} \phi_j \overline{\phi_j(w)} \right\|^2 = \sum_{j=N+1}^{M} |\phi_j(w)|^2.$$

If we can show that the sequence of complex numbers $\{\phi_j(w)\}$ is square summable, then it follows that the partial sums of (15) will be a Cauchy sequence in $A^2(\Omega)$.

The proof uses Proposition III.1.7, Exercise 7 from Chapter II, and the Cauchy-Schwarz inequality on l^2. Suppose $f \in A^2(\Omega)$. Then, by orthonormal expansion,

$$f = \sum \langle f, \phi_j \rangle \phi_j.$$

Since evaluation at z is continuous (Corollary III.1.9) we obtain

$$f(z) = \sum \langle f, \phi_j \rangle \phi_j(z).$$

Taking absolute values and using Corollary III.1.9 again gives

$$\left| \sum \langle f, \phi_j \rangle \phi_j(z) \right| = |f(z)| \le C_z \| f \|.$$

Exercise 7 from Chapter II shows that

$$\sum |\phi_j(z)|^2 \le \sup_{\| f \|=1} |f(z)|^2 \le C_z^2. \tag{16}$$

Thus the sequence of complex numbers $\{\phi_j(z)\}$ is square summable. Therefore the partial sums of (15) are a Cauchy sequence and hence converge to a limit in $A^2(\Omega)$.

Let K be a compact subset of Ω. Next we take the supremum over $z \in K$ of both sides of (16). By Proposition III.1.7 we obtain

$$\sup_{z \in K} \sum | \phi_j(z) |^2 \le C_K.$$

This inequality yields the uniform convergence on compact subsets of the diagonal (that is, where $z = w$) of (14).

Since the sequence $\{\phi_j(z)\}$ is square summable, we may use the Cauchy-Schwarz inequality on l^2 to estimate

$$| \sum \phi_j(z) \overline{\phi_j(w)} | \le \left(\sum | \phi_j(z) |^2 \right)^{\frac{1}{2}} \left(\sum | \phi_j(w) |^2 \right)^{\frac{1}{2}}.$$

Therefore uniform convergence on the diagonal implies uniform convergence on $\Omega \times \Omega^*$.

We have established the uniform convergence of the sum in (14) on compact subsets of $\Omega \times \Omega^*$, and the convergence (in the first variable) in $A^2(\Omega)$. Combining these results with Theorem III.1.5 yields property B2). We verify that the limit function is the Bergman kernel by using Lemma III.3.3. Thus it remains to show that the sum satisfies B1) and B3).

Since any partial sum $S(z, \overline{w})$ of (14) satisfies

$$S(z, \overline{w}) = \overline{S(w, \overline{z})},$$

and complex conjugation is continuous, the sum $B(z, \overline{w})$ also satisfies this conjugate symmetry property. Therefore B1) holds.

The proof of B3) is also easy. For $h \in A^2(\Omega)$ we have

$$h = \sum_j \langle h, \phi_j \rangle \phi_j$$

by orthonormal expansion. Since evaluation at z is continuous on $A^2(\Omega)$, the statement

$$h(z) = \sum_j \langle h, \phi_j \rangle \phi_j(z) \tag{17}$$

holds for all $z \in \Omega$. Writing the inner products in (17) as integrals, and then interchanging the order of integration and summation, gives B3). The interchange is valid because of the uniform convergence on compact subsets. See Exercise 13.

By Lemma III.3.3 there is a unique function, namely the Bergman kernel function, satisfying B1), B2), and B3). The sum in (14) must therefore be the Bergman kernel function, and it must be independent of the orthonormal system used. $\qquad\square$

The formula (14) for the Bergman kernel function involves an infinite sum and does not generally represent an elementary function. The series does have an explicit formula in a few cases; next we find one for the unit ball. This computation will be applied in Chapter VII.

Theorem III.3.5. The Bergman kernel function for the unit ball in $\mathbf{C}^n$ is given by

$$B(z, \bar{\zeta}) = \frac{n!}{\pi^n} \frac{1}{(1 - \langle z, \zeta \rangle)^{n+1}}. \tag{18}$$

Proof. Set $\phi_\alpha(z) = \frac{z^\alpha}{c_\alpha}$, where the c_α were computed in Proposition III.2.5. The $\{\phi_\alpha\}$ define a complete orthonormal system by Proposition III.2.5. By Proposition III.3.4

$$B(z, \bar{\zeta}) = \sum_\alpha \frac{z^\alpha \bar{\zeta}^\alpha}{c_\alpha^2} = \sum_\alpha \frac{(z\bar{\zeta})^\alpha}{c_\alpha^2}. \tag{19}$$

To verify that (19) yields (18), we will expand the right-hand side of (18).

For $|x| < 1$ we differentiate the geometric series n times to obtain

$$\frac{n!}{(1-x)^{n+1}} = \left(\frac{d}{dx}\right)^n \frac{1}{1-x}$$

$$= \left(\frac{d}{dx}\right)^n \sum_{k \geq 0} x^k = \sum_{k \geq 0} \frac{(n+k)!}{k!} x^k. \tag{20}$$

We will replace x with $\langle z, \zeta \rangle$ in (20). The multinomial expansion gives

$$\langle z, \zeta \rangle^k = \sum_{|\alpha|=k} \binom{k}{\alpha} (z\bar{\zeta})^\alpha = \sum_{|\alpha|=k} \frac{k!}{\alpha!} (z\bar{\zeta})^\alpha.$$

Plugging this formula into (20) gives

$$\frac{n!}{(1 - \langle z, \zeta \rangle)^{n+1}} = \sum_{\alpha} \frac{(n + |\alpha|)!}{\alpha!} (z\overline{\zeta})^{\alpha}.$$

Inserting the formula

$$\frac{1}{c_{\alpha}^2} = \frac{(n + |\alpha|)!}{\pi^n \alpha!}$$

gives

$$\frac{n!}{\pi^n (1 - \langle z, \zeta \rangle)^{n+1}} = \sum_{\alpha} \frac{(z\overline{\zeta})^{\alpha}}{c_{\alpha}^2}.$$

By Proposition III.3.4 the right-hand side is the Bergman kernel for the ball.   □

Remark III.3.6. This explicit formula serves as a model for general results about the Bergman kernel function. We give two examples here. When the second variable w is fixed, $B(z, \overline{w})$ is holomorphic in z. The explicit formula (for the unit ball) shows that this function has a holomorphic extension past the unit sphere. This property is crucial for establishing some deep results in the theory of holomorphic mappings. See [Kr]. Note also that the Cauchy-Schwarz inequality implies that the kernel function, for the unit ball $\mathbf{B}_n$, is smooth on the product of the closed unit ball with itself, except on the boundary diagonal. There is a general class of domains, called *pseudoconvex domains of finite type*, for which this regularity property holds. The proofs use sophisticated methods of partial differential equations rather than explicit knowledge of the kernel function. See [FK], [Kr], and [CD2] for more on the partial differential equations involved, and see [D] for the definition of pseudoconvex domain of finite type.

Remark III.3.7. Methods other than summing an orthonormal series sometimes work for computing the Bergman kernel function. For $\mathbf{B}_n$, there is an elegant computation using the automorphism group of the ball. We pause to describe this approach.

First we recall that a *holomorphic automorphism* of a domain Ω in $\mathbf{C}^n$ is a holomorphic mapping $\phi : \Omega \to \Omega$ that is injective and surjective. It follows that the inverse transformation ϕ^{-1} is also holomorphic. See page 29 of [D] for example. (The holomorphicity of the inverse is a complex variables phenomenon; by contrast the smooth mapping $x \to x^3$ on the real line is injective and surjective, but its inverse is not smooth.) The collection of holomorphic automorphisms of a bounded domain forms a Lie group. It is relatively easy to compute this group for the unit ball $\mathbf{B}_n$. See [Ru1] for a complete discussion. Each automorphism ϕ can be written $U h_a$, where U is a unitary transformation of $\mathbf{C}^n$, and h_a is a *linear fractional transformation* that satisfies $h_a(0) = a$ and $h_a(h_a(z)) = z$. For an appropriate linear transformation L_a we have

$$h_a(z) = \frac{a - L_a z}{1 - \langle z, a \rangle}.$$

From the definition of the Bergman kernel it is fairly easy (Exercise 15) to prove a change of variables formula. Let $h : \Omega \to \Omega$ be a holomorphic automorphism. Then

$$B(h(z), \overline{h(w)}) \det(dh(z)) \overline{\det(dh(w))} = B(z, \overline{w}). \qquad (21)$$

Let Ω be the unit ball $\mathbf{B}_n$. We apply the change of variables formula (21) when $z = w = 0$ and $h = h_a$; one computes $\det(dh(0))$ by elementary calculus. The value of $B(0, 0)$ is the reciprocal of the volume of the unit ball; this follows from (19) because all terms but the first in the expansion (19) vanish at the origin. We can thus determine $B(a, \overline{a})$ as follows:

$$B(a, \overline{a}) \mid \det(dh(0)) \mid^2 = B(h(0), \overline{h(0)}) \mid \det(dh(0)) \mid^2 = B(0, 0).$$

By polarization we can then find $B(a, \overline{b})$ for any points a and b in the unit ball.

We give the details when $n = 1$. Here

$$h_a(z) = \frac{a - z}{1 - z\overline{a}}$$

and hence $dh(0) = h'(0) = -1 + |a|^2$. Therefore

$$B(a, \overline{a})(1 - |a|^2)^2 = B(a, \overline{a}) | h'(0) |^2$$

$$= B(h(0), \overline{h(0)}) | h'(0) |^2 = B(0, 0) = \frac{1}{\pi}.$$

Hence $B(a, \overline{a}) = \frac{1}{\pi} \frac{1}{(1-|a|^2)^2}$, and by polarization we obtain $B(a, \overline{b}) = \frac{1}{\pi} \frac{1}{(1-a\overline{b})^2}$. In the same fashion we can obtain formula (18) for the Bergman kernel of $\mathbf{B}_n$.

Exercise 8. Both the definition of the Bergman kernel function and H4) from Example II.1.6 assume that Ω is a bounded domain. Why?

Exercise 9. Let p be a positive real number, and let Ω denote the domain in $\mathbf{C}^2$ defined by $\Omega = \{z : |z_1|^2 + |z_2|^{2p} < 1\}$. Show that the monomials are orthogonal, and compute their norms. Assuming that the normalized monomials form a complete orthonormal system for $A^2(\Omega)$, compute the Bergman kernel function for Ω by summing the series (14). See [D1] and [BFS] for generalizations.

Exercise 10. Suppose that $\{L_\alpha\}$ is a collection of vectors in a Hilbert space, indexed by the monomials in n variables. Formulate an analogue of Proposition II.5.1 using a generating function with $t \in \mathbf{C}^n$.

Exercise 11. Compute the n variable generating function (as in Exercise 10) for the collection of normalized monomials on $\mathbf{B}_n$.

Exercise 12. Find the relationship between the Bergman kernel function for the Cartesian product of two bounded domains and the Bergman kernel functions for the respective domains.

Exercise 13. Verify the validity of the interchange of infinite sum and integral in the last step of the proof of Proposition III.3.4.

Exercise 14. Suppose that f is holomorphic on $\mathbf{B}_1$, and hence represented by a convergent power series $\sum a_j z^j$. What is the condition on $\{a_j\}$ for f to be in $A^2(\mathbf{B}_1)$?

Exercise 15. Prove the change of variables formula (21) for the Bergman kernel function.

CHAPTER **IV**
Linear Transformations and Positivity Conditions

This book aims to provide a systematic discussion of inequalities that bear on complex analysis. Many of these come from linear algebra and positivity conditions for Hermitian linear transformations. We will prove the basic facts about linear transformations on complex Euclidean spaces, and then generalize them to results about bounded transformations on Hilbert spaces. Linear transformations on Hilbert spaces are also known as *operators*, and their study is often called *operator theory*.

We prove the spectral theorem for Hermitian operators in finite dimensions in this chapter. We defer to Chapter V its generalization to compact Hermitian operators on a Hilbert space. In this chapter we derive the required information about eigenvalues by combining analytic and algebraic reasoning; in particular we find eigenvalues via an optimization procedure. Because of our emphasis on inequalities, we discuss positive definite matrices in considerable detail. The results include several characterizations of positive definite matrices.

The chapter closes with two sections on elementary Fourier analysis. The material on Fourier series includes proofs of Hilbert's inequality, Fejér's theorem, and Herglotz's theorem. We state without proof Bochner's generalization of Herglotz's theorem to functions of positive type, and we also discuss Wirtinger's inequality.

IV.1   Adjoints and Hermitian forms

The main ideas in this section are adjoints, Hermitian forms, unitary operators, and positivity conditions for linear transformations. Our first order of business is the notion of adjoint of an operator. Adjoints will be crucial for us; the mapping sending an operator into its adjoint generalizes complex conjugation.

Two examples nicely illustrate this point of view. First consider the point of view in which a complex number w is identified with the linear transformation on $\mathbf{C}$ given by multiplication by w. Its adjoint is multiplication by $\overline{w}$. Next, for an operator L with adjoint L^*, the product L^*L is in a certain precise sense a *nonnegative real operator*, just as $\overline{z}z$ is a nonnegative real number.

Definition IV.1.1. Let $L : \mathcal{H} \to \mathcal{H}'$ be a bounded linear transformation between Hilbert spaces. The *adjoint* L^* of L is the linear transformation mapping $\mathcal{H}'$ to $\mathcal{H}$ defined by the formula

$$\langle Lz, w \rangle = \langle z, L^*w \rangle \tag{1}$$

for all z in $\mathcal{H}$ and all w in $\mathcal{H}'$.

Formula (1) requires some elaboration. For fixed w, the mapping sending z into $\langle Lz, w \rangle$ is a bounded linear functional on $\mathcal{H}$. By the Riesz lemma (Theorem II.2.4) there is a unique $\zeta \in \mathcal{H}$ such that $\langle Lz, w \rangle = \langle z, \zeta \rangle$ for all z. This ζ depends on w; let us call it L^*w. After taking complex conjugates of both sides of (1), we see that the mapping sending w into L^*w is linear. Hence there is a well-defined linear transformation L^* such that (1) holds.

In the finite-dimensional case we often identify a linear transformation with a matrix. Suppose $L : \mathbf{C}^n \to \mathbf{C}^N$. Let $\{e_j\}$ denote the standard basis in the domain, and let $\{\epsilon_j\}$ denote the standard basis in the range. The jk-th entry L_{jk} of the matrix of L is defined by $L_{jk} = \langle L(e_k), \epsilon_j \rangle$. Exercise 1 asks the reader to verify that the matrix of L^* is the conjugate transpose of the matrix of L. Its jk-th entry is $\overline{L_{kj}}$.

Exercise 1. Prove that the matrix of the adjoint of a linear transformation L in finite dimensions is the conjugate transpose of the matrix of L.

Exercise 2. Prove that $|| L^* || = || L ||$.

Definition IV.1.2. Let $L : \mathcal{H} \to \mathcal{H}$ be a bounded linear transformation.

1) L is *Hermitian* (or self-adjoint) if $L = L^*$.

2) L is *positive definite* if there is a constant $c > 0$ such that, for all $z \in \mathcal{H}$,

$$\langle Lz, z \rangle \geq c|| z ||^2.$$

3) L is *nonnegative definite* if, for all $z \in \mathcal{H}$,

$$\langle Lz, z \rangle \geq 0.$$

4) L is *unitary* if $L^*L = LL^* = I$; equivalently, if $L^* = L^{-1}$.

5) L is a *projection* if $L^2 = L$.

Example IV.1.3. The simplest example of a projection is given by the formula $L(z) = \langle z, w \rangle w$ where $|| w || = 1$. To see this first note that $L(w) = w$, and then compute

$$L(L(z)) = L(\langle z, w \rangle w) = \langle z, w \rangle L(w) = \langle z, w \rangle w = L(z).$$

This projection L is also Hermitian. To see this write

$$\langle Lz, \zeta \rangle = \langle z, w \rangle \langle w, \zeta \rangle = \langle z, \langle \zeta, w \rangle w \rangle = \langle z, L\zeta \rangle.$$

It follows that $L = L^*$. See also Proposition IV.1.4 below.
    Finally this L is also nonnegative definite, because

$$\langle Lz, z \rangle = | \langle z, w \rangle |^2 \geq 0.$$

We pause to observe some differences between the finite- and infinite-dimensional cases. Suppose first that $\mathcal{H}$ is finite-dimensional, $L:\mathcal{H} \to \mathcal{H}$ is linear, and M exists such that $ML = I$. Then $LM = I$ and $M = L^{-1}$. When $\mathcal{H}$ is infinite-dimensional this conclusion fails. See Example IV.1.8 below. There the operator L annihilates only the zero vector, yet L is not invertible. This example illustrates why part 4) of Definition IV.1.2 assumes that each of L^*L and LL^* is the identity operator.

Next we consider the definition of positive definite. In part 2) of Definition IV.1.2 we require a positive constant c such that, for all z,

$$\langle Lz, z \rangle \geq c\|z\|^2.$$

In the finite-dimensional case, the existence of such a c is equivalent to

$$\langle Lz, z \rangle > 0$$

for all nonzero z. The proof of equivalence relies on the compactness of the unit sphere. In the infinite-dimensional case these two conditions are not equivalent. We give a simple example.

Consider an infinite diagonal matrix whose n-th diagonal entry is $\frac{1}{n+1}$. To realize this matrix as an operator L, we can let $\mathcal{H} = A^2(\mathbf{B}_1)$. The monomials form a complete orthogonal system by Proposition III.2.5. For each $n \geq 0$ we put $L(z^n) = \frac{z^n}{n+1}$, and extend by linearity. Then $\langle Lf, f \rangle > 0$ for all nonzero f in $\mathcal{H}$; on the other hand L is not positive definite according to Definition IV.1.2. The distinction will matter in Application IV.5.5. See also Lemma VII.6.4 and Exercise 5 of Chapter VII.

We return to the general discussion. Let $L : \mathcal{H} \to \mathcal{H}$ be a linear transformation, and consider the function ϕ from $\mathcal{H} \times \mathcal{H}$ to $\mathbf{C}$ defined by

$$\phi(z, w) = \langle Lz, w \rangle. \tag{2}$$

This function ϕ is linear in the first variable and conjugate linear in the second, and thus ϕ defines a sesquilinear form. When L is Hermitian the additional property $\overline{\phi(z, w)} = \phi(w, z)$ holds; we call ϕ a *Hermitian form* on $\mathcal{H}$. Conversely, given a function $\phi : \mathcal{H} \times \mathcal{H} \to \mathbf{C}$ that is linear in the first variable and conjugate linear in the second, there is a linear transformation L such that (2) holds. Proposition IV.1.5 below shows that L is determined by ϕ; Exercise 3 asks the reader to provide an example (in the real case) where a quadratic form does not uniquely determine an underlying linear transformation.

Positive definite linear transformations play a major role in this book; they are always Hermitian. Our immediate aim is to establish this result, using the polarization identity from Theorem II.4.3. Proposition IV.1.5 gives a second application of polarization; the Hermitian form determines the (linear) operator. We first require a simple result; it gives a simple indication of a basic aspect of quantum mechanics. *Hermitian* operators behave something like real rather than complex numbers.

Proposition IV.1.4. Let L be a bounded linear transformation on a Hilbert space. The following three statements are equivalent:

1) L is Hermitian.
2) For all z and ζ,

$$\langle Lz, \zeta \rangle = \langle z, L\zeta \rangle = \overline{\langle L\zeta, z \rangle}.$$

3) $\langle Lz, z \rangle$ is real for all z.

Proof. By definition of adjoint, $\langle L^*z, \zeta \rangle = \langle z, L\zeta \rangle$ for all z, ζ. Therefore the first equality in 2) holds for all z, ζ if and only if L is Hermitian. The second equality in 2) always holds, by Definition II.1.1. Therefore 1) and 2) are equivalent. Also 2) implies 3) by setting $z = \zeta$.

It remains to prove that 3) implies 2). The third part of Theorem II.4.3 on polarization identities shows that the reality of $\langle Lv, v \rangle$ for all v implies the equality of the first and third terms in 2), and hence the equality of all three terms. Therefore 3) implies 2). $\square$

Proposition IV.1.5. Suppose A and B are bounded linear transformations on a Hilbert space and $\langle Az, z \rangle = \langle Bz, z \rangle$ for all z. Then $A = B$.

Proof. The operator $A - B$ satisfies $\langle (A - B)z, z \rangle = 0$ for all z. By (17) of Chapter II we have $\langle (A - B)z, w \rangle = 0$ for all z and w; hence $(A - B)z = 0$ for all z. Thus $A = B$. □

The reader should note that this conclusion fails for linear transformations on real vector spaces with inner products.

Exercise 3. Give an example of a linear transformation $L : \mathbf{R}^2 \to \mathbf{R}^2$ such that $\langle Lv, v \rangle = 0$ for all v but L is not the zero operator.

Exercise 4. Suppose that $L : \mathcal{H} \to \mathcal{H}$ is a Hermitian linear transformation. Simplify the expression

$$\langle L(z + w), z + w \rangle - \langle L(z - w), z - w \rangle.$$

Exercise 5. Let L be a linear transformation on an inner product space over the *real* numbers. What should the definition of L^* be? How does this affect Exercise 1? Next suppose that $\langle Lx, x \rangle \geq 0$ for all x. Show that L need not equal L^*. Explain the difference between the real and complex cases.

Exercise 6. Give an example of a projection that is not Hermitian.

Exercise 7. (Difficult) Let L be an everywhere defined linear transformation that satisfies

$$\langle Lz, w \rangle = \langle z, Lw \rangle$$

for all z and w. Prove that L must be bounded.

Remark IV.1.6. (A return to the Bergman kernel) The ideas in this chapter clarify the material on the Bergman kernel from Chapter III. Let $\mathcal{H} = L^2(\Omega)$. Since $A^2(\Omega)$ is a closed subspace of $\mathcal{H}$, there is a unique Hermitian projection $P : \mathcal{H} \to A^2(\Omega)$. For $g \in \mathcal{H}$ we see that Pg is the unique holomorphic function such that $\| Pg - g \| \leq \| h - g \|$ for all $h \in A^2(\Omega)$. This mapping P is linear and satisfies $P^2 = P^* = P$. It is called the Bergman projection; P is given by the formula

$$Pg(z) = \int_\Omega g(w) B(z, \overline{w}) \, dV(w)$$

where B is the Bergman kernel function. Property B1) of the Bergman kernel from Lemma III.3.2 is equivalent to the Hermitian symmetry of P.

The study of bounded operators includes a simple heuristic idea. Hermitian operators behave something like real numbers, positive operators behave something like positive numbers, and unitary operators behave something like complex numbers on the unit circle. We conclude this section by proving some simple facts about unitary operators, thereby providing a glimpse of how unitary operators behave like complex numbers of absolute value 1.

Proposition IV.1.7. Let $\mathcal{H}$ be a Hilbert space. The following three statements about a linear transformation $U : \mathcal{H} \to \mathcal{H}$ are equivalent:

U1) U is unitary.

U2) U is surjective and U preserves inner products; that is, for all z and w, we have

$$\langle Uz, Uw \rangle = \langle z, w \rangle.$$

U3) U is surjective and U preserves norms; that is, for all z,

$$\| Uz \|^2 = \| z \|^2.$$

When $\mathcal{H}$ is finite-dimensional, surjectivity in 2) and 3) is an unnecessary hypothesis.

Proof. First we claim that U preserves all inner products if and only if U preserves norms. This claim follows either from the polarization

identity (Proposition II.4.1), or by the following argument. One implication is trivial. For the non-trivial implication suppose U preserves
norms. Then

$$\langle z, z \rangle = \| z \|^2 = \| U z \|^2 = \langle U^* U z, z \rangle.$$

By Proposition IV.1.5, $U^* U$ must be the identity. Thus

$$\langle U z, U w \rangle = \langle U^* U z, w \rangle = \langle z, w \rangle$$

and U preserves inner products. Thus statements U2) and U3) are
equivalent.

Suppose next that statement U1) holds; thus $U^* = U^{-1}$. Since U
is invertible it is surjective. Furthermore, for all $z, w \in \mathcal{H}$, we have

$$0 = \langle (U^* U - I) z, w \rangle \tag{3}$$

and hence

$$\langle z, w \rangle = \langle U^* U z, w \rangle = \langle U z, U w \rangle.$$

Thus statement U2) holds.

Suppose conversely that statement U2) holds. Then (3) holds, and
$U^* U = I$. The formula $U^* U = I$ does not suffice to prove that U
is invertible. If U is surjective, however, then the existence of the left
inverse U^* implies that U^* is also a right inverse. To see this, choose
$w \in \mathcal{H}$. Then $w = U v$ for some v, and hence $U^* w = U^* U v = v$.
Applying U we obtain $U U^* w = U v = w$. Since w was arbitrary we
obtain $U U^* = I$.

In the finite-dimensional case, the hypothesis that U is surjective
is not needed. The left inverse is automatically a right inverse, and
hence U is surjective if it preserves inner products. $\square$

Example IV.1.8. Suppose $\mathcal{H} = l^2$, and define L by

$$L(a_1, a_2, \ldots) = (0, a_1, a_2, \ldots).$$

Then $L^*L = I$, but LL^* is not the identity. In fact L^* annihilates the sequence $(1, 0, 0, \ldots)$. Thus L preserves all inner products without being unitary.

Definition IV.1.9. Let $L: \mathcal{H} \to \mathcal{H}'$ be a linear transformation. Then L is an *isometry* if $\| Lz \| = \| z \|$ for all $z \in \mathcal{H}$.

By Proposition IV.1.7, a linear transformation from a finite-dimensional space to itself is unitary if and only if it is an isometry. The operator L from Example IV.1.8 is an isometry of an infinite-dimensional space that is not unitary. The map $z \to (0, z)$ from $\mathbf{C}$ to $\mathbf{C}^2$ is an isometry; the analogous mapping in infinite dimensions can be regarded as a mapping from a space to itself.

IV.2 Solving linear equations

Solving linear equations surely qualifies as one of the dominant issues in the history of mathematics. We next consider some of the particular issues that arise when considering linear equations in the Hilbert space setting.

The *nullspace* of a bounded linear transformation $L : \mathcal{H} \to \mathcal{H}'$, written $N(L)$, consists of those $z \in \mathcal{H}$ such that $Lz = 0$. It is a *closed* linear subspace of $\mathcal{H}$ because L is continuous. Another important subspace is the *range* of L, written $R(L)$. It consists of those $w \in \mathcal{H}'$ for which there is some $z \in \mathcal{H}$ with $Lz = w$. Then $R(L)$ is a subspace of $\mathcal{H}'$ that need not be closed.

Example IV.2.1. Suppose that $\mathcal{H} = \mathcal{H}' = l^2$. Define L by

$$L(a_1, a_2, \ldots, a_j, \ldots) = \left(a_1, \frac{a_2}{2}, \ldots, \frac{a_j}{j}, \ldots \right).$$

It is clear that L is linear, that $L = L^*$, and that $N(L) = \{0\}$. We will show that $R(L)$ is not closed. Let z_n denote the element of $\mathcal{H}$

whose terms are $1, \frac{1}{2}, \ldots, \frac{1}{j}, \ldots, \frac{1}{n}, 0, 0, \ldots$. The sequence $\{z_n\}$ obviously converges in $\mathcal{H}$. Each z_n is in $R(L)$, but the limit is not.

The next result and the subsequent discussion indicates why it so useful for an operator to have closed range.

Proposition IV.2.2. Let $L : \mathcal{H} \to \mathcal{H}'$ be a bounded linear transformation. The following four statements hold:

1) $N(L)$ and $R(L^*)$ are orthogonal.
2) $N(L) \oplus R(L^*)$ is dense in $\mathcal{H}$.
3) If $R(L^*)$ is closed, then $N(L) \oplus R(L^*) = \mathcal{H}$.
4) If $R(L)$ is closed, then $N(L^*) \oplus R(L) = \mathcal{H}'$.

Proof. Suppose that $u \in N(L)$ and $v \in R(L^*)$. Then $v = L^*w$ for some w, and

$$\langle u, v \rangle = \langle u, L^*w \rangle = \langle Lu, v \rangle = 0.$$

Therefore 1) is true. To prove 2), it suffices to show that if z is orthogonal to both subspaces, then $z = 0$. Given such a z we have $0 = \langle z, L^*w \rangle$ for all w. Hence $0 = \langle Lz, w \rangle$ for all w, and therefore $Lz = 0$. Thus $z \in N(L)$, and also z is orthogonal to $N(L)$. Hence $0 = \langle z, z \rangle = \|z\|^2$, so $z = 0$.

Since $N(L)$ is closed, and the orthogonal sum of closed subspaces is closed, 3) follows from 2). To prove 4), we interchange the roles of L and L^* in 3). Since $(L^*)^* = L$, we obtain 4). $\qquad\square$

Proposition IV.2.2 clarifies Example IV.2.1. In this case $L = L^*$. If $R(L)$ were closed, then Proposition IV.2.2 would guarantee that $\mathcal{H} = N(L) \oplus R(L^*) = N(L) \oplus R(L)$. Since $N(L) = \{0\}$, we see that L would have to be surjective. It is clear why the image of L cannot be everything! The condition $\sum |a_j|^2 < \infty$ is weaker than the condition $\sum j^2 |a_j|^2 < \infty$.

Remark IV.2.3. Example IV.2.1 illustrates a general idea. The operator L is a type of smoothing operator. Suppose $a \in \mathcal{H}$, and the sequence $\{a_j\}$ contains infinitely many nonzero terms. Then La is *smoother* than a, in the sense that its terms tend to zero more rapidly. Suppose the a_j are the coefficients in an orthonormal expansion $\sum_{j=1}^{\infty} a_j e^{2\pi ijx}$ for some $f \in L^2[0, 1]$. Let Lf denote the function given by $\sum_{j=1}^{\infty} \frac{a_j}{j} e^{2\pi ijx}$. Then Lf is actually smoother than f in the sense of calculus.

These considerations of null spaces and ranges are particularly valuable in applied mathematics, where one needs to solve linear integral equations. We discuss one of the main points now. Suppose that $L : \mathcal{H} \rightarrow \mathcal{H}'$ and we wish to solve the linear equation $Lu = w$. A solution need not exist; for existence w must be in $R(L)$. If it is, then a solution exists. The solution will be unique only when $N(L)$ consists of 0 alone. When $R(L)$ is closed, there is a simple test for whether w is in $R(L)$; by 4) of Proposition IV.2.2, w must be orthogonal to $N(L^*)$.

This statement and various reformulations of it comprise part of what is called the Fredholm alternative. We repeat it here.

Proposition IV.2.4. Suppose that $L : \mathcal{H} \rightarrow \mathcal{H}'$ and $R(L)$ is closed. We can solve $Lu = w$ if and only if w is orthogonal to $N(L^*)$.

What do we do when w is not in the range of L? Since the equation $Lu = w$ has no solution, we instead seek an approximate solution by finding the u_1 that minimizes $|| Lu - w ||^2$ over all $u \in \mathcal{H}$. This approach is called the *method of least squares*. Before stating the theorem about least squares, we note the following simple fact.

Lemma IV.2.5. Suppose $L : \mathcal{H} \rightarrow \mathcal{H}'$ is linear. Then $N(L) = N(L^*L)$.

Proof. If $Lv = 0$, then $L^*Lv = 0$, so $N(L) \subset N(L^*L)$. On the other hand, if $L^*Lv = 0$, then

$$0 = \langle L^*Lv, v \rangle = \| Lv \|^2,$$

so $Lv = 0$. Therefore $N(L^*L) \subset N(L)$. Thus $N(L) = N(L^*L)$. $\square$

We can now establish the basic result about approximate solutions to linear equations. Statement 1) in Theorem IV.2.6 gives the existence of an approximate solution. Statement 2) tells us how to find it, and statement 3) tells us when the solution is unique.

Theorem IV.2.6. (Least Squares) Suppose $L : \mathcal{H} \to \mathcal{H}'$ and $R(L)$ is closed. The following three statements then hold:

1) For each $w \in \mathcal{H}'$, there exists u_1 in $\mathcal{H}$ for which $\| Lu_1 - w \|^2$ achieves the minimum value of $\| Lu - w \|^2$ for $u \in \mathcal{H}$.
2) Such a u_1 satisfies the equation $L^*Lu_1 = L^*w$.
3) The operator L^*L is injective if and only if $N(L) = \{0\}$.

Proof. Choose $w \in \mathcal{H}'$. Since $R(L)$ is closed, Theorem II.2.1 guarantees that there is a unique $z \in R(L)$ that minimizes $\| z - w \|$. Since $z = Lu_1$ for some (not necessarily unique) u_1, statement 1) follows.

By Corollary II.2.2 we may write $w = Lu_1 \oplus \zeta$ where ζ is orthogonal to $R(L)$. Note that Lu_1 and ζ are uniquely determined. By 4) of Proposition IV.2.2, $\zeta \in N(L^*)$. Therefore

$$L^*(w) = L^*(Lu_1 + \zeta) = L^*Lu_1 + L^*\zeta = L^*Lu_1,$$

and hence statement 2) holds.

Statement 3) holds because Lemma IV.2.5 guarantees that $N(L^*L) = N(L)$. $\square$

Exercise 8. (Application) Suppose we are given n pairs of real numbers (x_j, y_j), thought of as data points. Assume that $n > k$, and that we seek a real polynomial p of degree (at most) k that best fits these points, in the least squares sense. Thus we wish to minimize the sum

$$\sum_{j=1}^{n}(y_j - p(x_j))^2$$

over all real polynomials p of degree at most k. Develop formulas for the minimizer in case $k = 1$ and $k = 2$. Suggestion: Use Theorem IV.2.6.

IV.3   Linearization

Minimization techniques help us understand orthogonal projections. It is natural to ask whether we can use the ideas of calculus to solve the minimization problem in Theorem IV.2.6. The answer is yes. Furthermore, optimization techniques play a key role throughout linear algebra. We next discuss linearization, the basic idea of differential calculus.

Differential calculus is the study of linear approximation. Let V and W be normed vector spaces over the real numbers. We write both norms as $\| \ \|$; this notation will not cause confusion here. Assume that V and W are Banach spaces; in particular they are complete metric spaces. Let Ω be an open subset of V, suppose $p \in \Omega$, and let $f : \Omega \to W$. Then f is differentiable at p if it is approximately linear. More precisely we have the following fundamental definition of the derivative.

Definition IV.3.1. Suppose $f : \Omega \to W$. We say that f is *differentiable* at p if there is a linear transformation $L : V \to W$ such that

$$f(p + h) = f(p) + L(h) + E(p, h), \qquad (4)$$

where the error term $E(p, h)$ defined by (4) must satisfy

$$\lim_{h \to 0} \frac{\| E(p, h) \|}{\| h \|} = 0.$$

This formula implies that L, when it exists, is unique. See Exercise 9. We write $df(p) = L$, and call $df(p)$ the derivative of f at p. In (4) the vector h must be sufficiently small so that $f(p + h)$ is defined.

We call $df(p)(h)$ the linear part of $f(p + h) - f(p)$. Definition IV.3.1 allows us to approximate $f(x)$ by $f(p) + df(p)(x - p)$ near p. A fundamental consequence is the chain rule: the composition of differentiable functions is differentiable, and the derivative of the composition is the composition of the derivatives. See Exercise 10.

Exercise 9. Prove that the linear transformation defined in Definition IV.3.1 is unique.

Exercise 10. Formulate and prove the chain rule for composition of differentiable mappings between Banach spaces.

Exercise 11. Let V be the normed vector space of bounded linear transformations from a Hilbert (or Banach) space to itself, with the usual norm. Let Ω denote the invertible elements of V. Define $f : \Omega \to V$ by $f(T) = T^{-1}$.

1) Prove that Ω is an open subset of V.
2) Find df.

Complex variable analogues of the derivative are useful in the following setting. Let $r : \mathbf{C}^n \times \mathbf{C}^n \to \mathbf{C}$ be a holomorphic function; setting the second variable equal to the conjugate of the first determines a function $z \to r(z, \overline{z})$. We naturally treat z and $\overline{z}$ as independent variables, and therefore require separate symbols for the derivatives with respect to z and to $\overline{z}$.

We consider the linear approximation to r in the z variable, and write

$$\frac{\partial r}{\partial z}(z, \overline{z})$$

for the linear part of

$$r(z + h, \overline{z}) - r(z, \overline{z}).$$

Similar notation and discussion applies for the derivative with respect to $\overline{z}$.

As an illustration consider the function r defined by $r(z, \overline{z}) = \langle Lz, z \rangle$, where L is a Hermitian linear transformation. By Proposition IV.1.4, r is real-valued. It is important to treat the z on the right-hand side of the inner product as a $\overline{z}$. As a function of z, r is differentiable, and $\frac{\partial r}{\partial z}$ is the linear map defined by $\frac{\partial r}{\partial z} h = \langle h, Lz \rangle$. To verify this statement we treat z and $\overline{z}$ as independent, and note that

$$\langle L(z + h), z \rangle = \langle Lz, z \rangle + \langle Lh, z \rangle.$$

Since $\langle Lh, z \rangle = \langle h, Lz \rangle$ we see that $\frac{\partial r}{\partial z}$ is the linear functional sending h to $\langle h, Lz \rangle$. Thus we have

$$\frac{\partial r}{\partial z}(h) = \langle h, Lz \rangle. \tag{5}$$

A warning seems appropriate; the term Lz appears as the right-hand entry in the inner product, and therefore carries a complex conjugate with it. In one variable, for example, $\frac{\partial}{\partial z} |z|^2 = \overline{z}$. The analogous formula in more variables is

$$\frac{\partial}{\partial z} \|z\|^2 = \langle \cdot, z \rangle;$$

this notation disguises the conjugation.

Application IV.3.2. We revisit the method of least squares. We wish to minimize $\|Lu - w\|^2$ over u. Assuming such a minimum exists, we try to find it by calculus. Define f by

$$f(u, \overline{u}) = \langle Lu - w, Lu - w \rangle.$$

Treating u and $\overline{u}$ as independent gives

$$f(u + h, \overline{u}) - f(u, \overline{u}) = \langle Lh, Lu - w \rangle = \langle h, L^*Lu - L^*w \rangle.$$

Thus the derivative $\frac{\partial f}{\partial u}$ is the linear functional defined by taking the inner product with $L^*Lu - L^*w$. Its vanishing gives the least squares formula from Theorem IV.2.6.   □

We continue with ideas from differential calculus by recalling the method of Lagrange multipliers. Let f be a smooth real-valued function on $\mathbf{R}^n$, called the *objective* function. Assume $1 \leq k < n$ and let $g_1, \ldots, g_k$ be smooth functions on $\mathbf{R}^n$. We wish to maximize (or minimize) $f(x)$ subject to the constraints that

$$g_1(x) = g_2(x) = \cdots = g_k(x) = 0.$$

Then all candidates for points where f achieves its constrained maximum (or minimum) can be found by considering those points x that satisfy the constraints and also for which either

1) $df(x)$ is a linear combination of the $dg_j(x)$, or

2) the $dg_j(x)$ are linearly dependent.

There is a corresponding result when the variables are complex. We will use the result only when $f(z, \overline{z})$ and $g_j(z, \overline{z})$ are Hermitian forms. Since they are real-valued, it suffices to take the differential with respect to the z variables. We can then replace 1) and 2) by

1c) $\frac{\partial f}{\partial z}(z, \overline{z})$ is a linear combination of the $\frac{\partial g_j}{\partial z}(z, \overline{z})$, or

2c) the $\frac{\partial g_j}{\partial z}(z, \overline{z})$ are linearly dependent.

Application IV.3.3. Suppose L is Hermitian, and consider the problem of maximizing $\langle Lz, z \rangle$ on the unit sphere in $\mathbf{C}^n$. Put $r(z) = \langle Lz, z \rangle$. The unit sphere is defined by the single constraint $g(z) = \langle z, z \rangle - 1$. Note that $\frac{\partial g}{\partial z}(z)$ is the linear functional given by taking the inner product with z. The method of Lagrange multipliers works nicely here. From formula (5) for the derivative of r, we see that the candidates for points where the maximum occurs are those unit vectors z for which Lz

is a multiple of z. Note that if $Lz = \lambda z$, and $\|z\| = 1$, then $\langle Lz, z \rangle = \lambda$. We study this optimization problem fully in the next section.

IV.4 Eigenvalues and the spectral theorem in finite dimensions

Our primary goal in this section is to prove the Spectral Theorem for Hermitian linear transformations on finite-dimensional Hilbert spaces. The word "spectral" means "relating to, or made by, a spectrum." For reasons coming from physics, the collection of eigenvalues of a linear transformation is known as its *point spectrum*. See Definition IV.4.1.

Even in the finite-dimensional case, the spectral theorem has many applications. It provides a nice way to define continuous functions of matrices, and it allows us to derive what we need to know about positive definite matrices. We recall the notion of eigenvalue.

Definition IV.4.1. Let $T : V \to V$ be a linear transformation on a vector space V. Suppose there is a nonzero vector z and a scalar λ such that $Tz = \lambda z$. Then z is called an *eigenvector* for T, and λ is called an *eigenvalue* for T. We also say that λ is the eigenvalue corresponding to the eigenvector z. The set of all eigenvalues is called the *point spectrum* of T. In case V is a space of functions, we sometimes say *eigenfunction* instead of eigenvector.

It is natural to identify the complex number λ with the linear transformation given by scalar multiplication by λ. In case λ is an eigenvalue for T, the linear transformation $\lambda - T$ is evidently not invertible. When V is finite-dimensional, this situation is the only way $(\lambda - T)^{-1}$ can fail to exist. When V is infinite-dimensional, things are more complicated. Even when λ is not an eigenvalue for T, the operator $\lambda - T$ need not have an inverse. In this case one says that λ is in the *spectrum* of T, but not in its *point spectrum*. In this book only the point spectrum arises.

Our next goal is to show that Hermitian linear transformations on $\mathbf{C}^n$ always have enough eigenvectors to span $\mathbf{C}^n$. Let L be a Hermitian linear transformation on $\mathbf{C}^n$. The function ϕ given by $\phi(z) = \langle Lz, z \rangle$ is real-valued and continuous. By the Min-Max Theorem from the Appendix, its restriction to any compact subset of its domain achieves its supremum value. Hence, on the unit sphere, there is some z at which $\phi(z)$ is maximal. By Application IV.3.3, z is an eigenvalue for L. Furthermore, since $\| z \|^2 = 1$, we have

$$\phi(z) = \langle Lz, z \rangle = \langle \lambda z, z \rangle = \lambda.$$

Therefore the maximum value of ϕ on the unit sphere is the maximum eigenvalue of L. Our next theorem generalizes this discussion to show that L has a basis of eigenvectors, and that each eigenvalue is the solution of an optimization problem.

Theorem IV.4.2. (**Spectral Theorem in finite dimensions**) Let V be a Hilbert space of dimension n and suppose $L : V \to V$ is a Hermitian linear transformation. The following statements hold:

1) There is an orthonormal basis of V consisting of eigenvectors $v_1, \ldots, v_n$ of L. The corresponding eigenvalues $\lambda_1, \ldots, \lambda_n$ are real.
2) For all $z \in V$, we have

$$\langle Lz, z \rangle = \sum_{j=1}^{n} \lambda_j | \langle z, v_j \rangle |^2. \tag{6}$$

3) For all $z \in V$, we have

$$Lz = \sum_{j=1}^{n} \lambda_j \langle z, v_j \rangle v_j. \tag{7}$$

Proof. We continue using the notation from the discussion preceding the statement of the theorem. There we showed that L has at least one eigenvector v_1 with corresponding real eigenvalue λ_1. This forms the basis step for a proof by induction. Suppose we have found eigenvectors $v_1, \ldots, v_k$ for $1 \le k < n$, these are orthogonal, and they have unit length. We consider the optimization problem of finding the maximum of $\phi(z)$ subject to the constraints 1) and 2):

1) $\| z \|^2 = 1$.

2) z is orthogonal to each of the vectors $v_1, \ldots, v_k$.

Again we wish to maximize a continuous function on a compact set (the unit sphere in a lower dimensional space); the maximum exists by the Min-Max Theorem. Again we use the Lagrange multiplier method to find where it occurs. The constraints are the vanishing of the real-valued functions g_j for $1 \le j \le k$ given by $g_j(z) = |\langle z, v_j \rangle|^2$ and g_{k+1} given by $g_{k+1}(z, \overline{z}) = \| z \|^2 - 1$. For $1 \le j \le k$ we have

$$\frac{\partial g_j}{\partial z}(z)(h) = \langle h, v_j \rangle \, \langle v_j, z \rangle = \langle h, \langle z, v_j \rangle v_j \rangle.$$

Thus the differential dg_j at z is the linear functional given by taking the inner product with $\langle z, v_j \rangle v_j$. These differentials are dependent only when some $\langle z, v_j \rangle$ vanishes; thus 2c) adds no new candidates.

The method of Lagrange multipliers therefore yields the equation

$$Lz = \sum \lambda_j \langle z, v_j \rangle v_j + \mu z.$$

The constraints that z is orthogonal to each v_j then imply $Lz = \mu z$. Therefore $\langle Lz, z \rangle = \langle \mu z, z \rangle = \mu$, because $\| z \|^2 = 1$. Thus we have obtained a unit vector z, which is also an eigenvector for L, and which is orthogonal to the previously found eigenvectors. By induction we therefore obtain n mutually orthogonal unit eigenvectors, where n is the dimension of V over $\mathbf{C}$. Hence the first conclusion of the theorem is true.

The second and third statements follow easily by orthonormal expansion. For $z \in V$ we write, via orthonormal expansion,

$$z = \sum_{j=1}^{n} \langle z, v_j \rangle v_j.$$

Applying L to both sides gives (7), because $L(v_j) = \lambda_j v_j$. Recall also that $\langle v_j, v_k \rangle = 0$ for $k \neq j$. Hence

$$\langle Lz, z \rangle = \left\langle \sum_{j=1}^{n} \langle z, v_j \rangle \lambda_j v_j, \sum_{k=1}^{n} \langle z, v_k \rangle v_k \right\rangle = \sum_{j=1}^{n} \lambda_j |\langle z, v_j \rangle|^2.$$

This proves (6) and completes the proof of all three statements.   □

The following simple corollary provides another example indicating a sense in which z and $\bar{z}$ are independent variables. A formal argument yields (8) by treating the z and $\bar{z}$ on the right-hand side of (6) as independent; a rigorous argument uses (7).

Corollary IV.4.3. Suppose that $L : V \to V$ is Hermitian. For all $z, w \in V$, we have

$$\langle Lz, w \rangle = \sum_{j=1}^{n} \lambda_j \langle z, v_j \rangle \langle v_j, w \rangle. \tag{8}$$

Proof. The conclusion follows immediately from (7) and orthonormality.   □

We next briefly discuss one application of Corollary IV.4.3. It is often useful to define functions of a linear transformation L. It is clear what we mean by powers of L, and hence also what we mean by $p(L)$ when p is a polynomial. It is also evident that if λ is an eigenvalue for L, then λ^k is an eigenvalue for L^k. These considerations suggest how to define $f(L)$ when f is continuous. The following definition provides the precise answer and a beautiful application of the spectral theorem.

Definition IV.4.4. Let L be Hermitian, and let X be a compact subset of $\mathbf{R}$ containing the spectrum of L. Suppose f is a continuous function on X. We define the linear transformation $f(L)$ by

$$f(L)(z) = \sum_{j=1}^{n} f(\lambda_j)\langle z, v_j \rangle v_j. \tag{9}$$

In (9) we are using the notation from Theorem IV.4.2.

Remark IV.4.5. The reader may wonder why Definition IV.4.4 is reasonable. What do we mean by $f(L)$ in the first place? The operator $f(L)$ has an unambiguous interpretation when f is a polynomial for example. According to Theorem IV.4.2, formula (9) is a correct statement when f is the identity function. It is easy to verify the correctness of (9) when $f(x)$ is a power of x; it follows that (9) holds for polynomials in x. By the Weierstrass approximation theorem ([F], [TBB], or Exercise 29), every continuous function on a compact subset of **R** is the uniform limit there of polynomials, so (9) is the natural formula for continuous functions as well. $\quad\square$

Exercise 12. (**Application**) Suppose that L is a linear transformation for which (7) holds, but we allow the eigenvalues λ_j to be complex. Assume that $|\lambda_j| \leq 1$ for each eigenvalue λ_j. Let $G_N(L)$ denote the linear transformation defined by

$$G_N(L) = \frac{1}{N} \sum_{j=0}^{N-1} L^j.$$

Determine $\lim_{N \to \infty} G_N(L)$. Suggestion: Use (11) from Chapter I and (7), (9).

We summarize the discussion about Hermitian operators on finite-dimensional Hilbert spaces. There is an orthonormal basis of eigenvectors for a Hermitian operator, and the corresponding eigenvalues are real. The proof considers the values of the Hermitian form $\langle Lz, z \rangle$ on the unit sphere. The maximum eigenvalue is the maximum value of this form on the unit sphere, and the minimum eigenvalue is the minimum value of this form on the unit sphere. The other eigenvalues are also obtained via constrained optimization.

IV.5   Positive definite linear transformations in finite dimensions

Our first corollary of the spectral theorem is a test for positive definiteness. We then discuss various alternative tests for positive definiteness.

Corollary IV.5.1. A Hermitian linear transformation $L : \mathbf{C}^n \to \mathbf{C}^n$ is positive definite if and only if all its eigenvalues are positive. The minimum eigenvalue is the largest c we may use in the defining condition for positive definiteness (Definition IV.1.2).

Proof. One implication is easy. If L is positive definite, and $Lz = \lambda z$, then we have

$$0 \leq c\|z\|^2 \leq \langle Lz, z \rangle = \lambda\|z\|^2;$$

hence $0 < c \leq \lambda$ for each eigenvalue λ.

Note incidentally that this implication does not require the assumption that L is Hermitian; by Proposition IV.1.4 a positive definite L must be Hermitian.

Conversely, suppose that all the eigenvalues are positive. This implication does require that L be Hermitian. Let λ_n denote the minimum eigenvalue, and let $v_1, \ldots, v_n$ be the orthonormal basis guaranteed by Theorem IV.4.2. We then have

$$\|z\|^2 = \left\| \sum_{j=1}^n \langle z, v_j \rangle v_j \right\|^2 = \sum_{j=1}^n |\langle z, v_j \rangle|^2.$$

Using (6) shows

$$\langle Lz, z \rangle = \sum_{j=1}^n \lambda_j |\langle z, v_j \rangle|^2 \geq \lambda_n \sum_{j=1}^n |\langle z, v_j \rangle|^2 = \lambda_n\|z\|^2,$$

and therefore L is positive definite. Thus we may choose $c = \lambda_n$ in the defining condition $\langle Lz, z \rangle \geq c\|z\|^2$ from Definition IV.1.2. We have already shown that c is at most λ_n. Combining these statements about

c proves that the minimum eigenvalue is the largest c we may choose in the defining condition for positive definiteness. $\square$

We next derive the standard tests for positive definiteness of an n-by-n Hermitian matrix of complex numbers.

Theorem IV.5.2. Let $L : \mathbf{C}^n \to \mathbf{C}^n$ be a linear transformation. The following three statements hold:

1) L is nonnegative definite if and only if there is a linear transformation $T:\mathbf{C}^n \to \mathbf{C}^n$ such that $L = T^*T$.

2) L is positive definite if and only if there is an *invertible* linear transformation $T : \mathbf{C}^n \to \mathbf{C}^n$ such that $L = T^*T$.

3) Let L_{kj} denote the entries of the matrix of L with respect to the standard basis of $\mathbf{C}^n$. Then L is positive definite if and only if there is a basis $\phi_1, \ldots, \phi_n$ of $\mathbf{C}^n$ such that

$$L_{kj} = \langle \phi_j, \phi_k \rangle.$$

Proof. Suppose first that $L = T^*T$. Then

$$\langle Lz, z \rangle = \langle T^*Tz, z \rangle = \| Tz \|^2 \geq 0. \tag{10}$$

Inequality (10) shows that L is nonnegative definite. Suppose T is also invertible. Define c by $\frac{1}{c} = \| T^{-1} \|$; then (10) can be sharpened:

$$\langle Lz, z \rangle = \| Tz \|^2 \geq c^2 \| z \|^2.$$

Therefore L is positive definite. We have proved one implication in each of 1) and 2).

Suppose L is nonnegative definite; then L is Hermitian and its eigenvalues are nonnegative. We construct T in 1) and 2) as a square root of L. Let $\{v_j\}$ be the orthonormal basis guaranteed by Theorem IV.4.2. Following Definition IV.4.4 we define the square root $\sqrt{L}$ on each basis element $\{v_j\}$ by

$$\sqrt{L}\,(v_j) = \sqrt{\lambda_j}\,v_j,$$

and extend the definition to $\mathbf{C}^n$ by linearity. Since the eigenvalues of $\sqrt{L}$ are real, it is Hermitian. The other implication in each of 1) and 2) now holds with $T = \sqrt{L}$.

Now we prove 3). First assume L is positive definite. Given the standard basis $\{e_j\}$, we define ϕ_j by $\phi_j = \sqrt{L}\,e_j$. Since L is positive definite, it is invertible, and $\sqrt{L}$ is also invertible. Therefore $\{\phi_j\}$ forms a basis for V. Furthermore, $\sqrt{L}$ is positive definite and hence Hermitian. Therefore we have

$$L_{kj} = \langle Le_j, e_k \rangle = \langle \sqrt{L}\,e_j, \sqrt{L}\,e_k \rangle = \langle \phi_j, \phi_k \rangle,$$

and one implication in 3) holds. Conversely suppose such a basis exists. Then

$$\langle Lz, z \rangle = \sum_{j,k=1}^{n} L_{kj} z_j \bar{z}_k = \sum_{j,k=1}^{n} \langle \phi_j, \phi_k \rangle z_j \bar{z}_k$$

$$= \left\|\sum_{j=1}^{n} z_j \phi_j\right\|^2 \geq c\|z\|^2. \tag{11}$$

The last step follows because the ϕ_j are linearly independent; we may choose c to be the (necessarily positive) minimum of $\|\sum_{j=1}^{n} z_j \phi_j\|^2$ on the unit sphere in $\mathbf{C}^n$. $\qquad\square$

Theorem IV.5.2 includes a characterization of positive-definiteness in terms of the entries of the matrix of a linear transformation with respect to the usual basis. We may apply this criterion to a matrix without thinking of it as the matrix of a linear transformation. An n-by-n matrix (M_{kj}) of complex numbers is positive definite if and only if there are n linearly independent vectors $\phi_1, \ldots, \phi_n$ such that $M_{kj} = \langle \phi_j, \phi_k \rangle$.

Remark IV.5.3. Sometimes it is convenient to rephrase one piece of Theorem IV.5.2 in the following way. Suppose that $\phi_1, \ldots, \phi_n$ are linearly independent elements in a Hilbert space $\mathcal{H}$. Let $L_{kj} = \langle \phi_j, \phi_k \rangle$. Then the Hermitian form on $\mathbf{C}^n$ defined by

$$\sum_{j,k=1}^{n} L_{kj} z_j \bar{z}_k$$

is positive definite.

Remark IV.5.4. Suppose L is Hermitian and we have shown

$$\langle Lz, z \rangle = \left\| \sum_{j=1}^{n} z_j \phi_j \right\|^2$$

for not necessarily linearly independent vectors $\{\phi_j\}$. The last step in (11) fails when these vectors are dependent. In this case the minimum value of $\| \sum z_j \phi_j \|^2$ on the unit sphere is zero. The reasoning in the proof shows that L is nonnegative definite if and only if there are vectors $\{\phi_j\}$, not necessarily independent, such that $L_{kj} = \langle \phi_j, \phi_k \rangle$. When they are independent, L is positive definite.

Theorem IV.5.2 has some surprising consequences.

Application IV.5.5. Let $\{x_j\}$ be a finite collection of distinct positive numbers. The matrix whose jk-th entry is $\frac{1}{x_j + x_k}$ is positive definite.

Proof. Let $\mathcal{H} = L^2([0, \infty), dx)$. The functions $t \to e^{-x_j t}$ are linearly independent in $\mathcal{H}$. The inner product of two of these functions is

$$\int_0^{\infty} e^{-(x_j + x_k)t} \, dt = \frac{1}{x_j + x_k},$$

so the conclusion follows from Remark IV.5.3.

In Application IV.5.5 we assumed that $\{x_j\}$ is a finite set; things are more subtle when $\{x_j\}$ is a countably infinite set. One can easily prove that the corresponding matrix is nonnegative definite; it is not necessarily positive definite in the sense of Definition IV.1.2. The interested reader should work out an infinite-dimensional analogue of Application IV.5.5.

Exercise 13. Prove that the functions $t \to e^{-x_j t}$ are linearly independent in $\mathcal{H}$.

Exercise 14. Prove the result in Application IV.5.5 by computing the principal minor determinants and using Theorem IV.5.7 below.

Generalization of Application IV.5.5. Let dg be a positive measure, and let

$$\mathcal{H} = L^2([0, \infty), \, dg).$$

Define a function f by the formula

$$f(x) = \int_0^\infty e^{-xt} \, dg(t).$$

Then, with x_j as in Application IV.5.5, $f(x_j + x_k)$ defines a positive definite matrix. We consider this idea again in Section 7.

There is a useful test for positive definiteness of a Hermitian matrix, involving minor determinants.

Definition IV.5.6. Suppose that A is an n-by-n matrix. For each j with $1 \leq j \leq n$, the j-th *leading principal minor determinant* of A is the determinant of the matrix A' whose entries are A_{kl} for $1 \leq k \leq j$ and $1 \leq l \leq j$. We call A' a *leading principal submatrix* of A.

Our next goal is to prove the following result.

Theorem IV.5.7. An n-by-n Hermitian matrix A is positive definite if and only if each of its leading principal minor determinants is a positive number.

We first discuss leading principal minor determinants. It is convenient to put $p_0 = 1$, and then to let p_k be the k-th leading principal minor determinant. Since A is Hermitian, each p_k is a real number. The following lemma will be used both to provide a key step in the proof of Theorem IV.5.7 and then to establish a famous inequality due to Hadamard.

Lemma IV.5.8. Let (A_{jk}) be an n-by-n Hermitian matrix, and suppose that $p_j \neq 0$ for $1 \leq j \leq n$. We then have

$$A_{nn} = \frac{p_n}{p_{n-1}} + \sum_{k=1}^{n-1} \frac{p_{k-1}}{p_k} | A_{nk} |^2. \tag{12}$$

Proof. We leave the details as an exercise. Here is one possible method. First solve (12) for p_n. Then show that we may assume that the leading $(n-1)$-by-$(n-1)$ block is diagonal. Finally use row operations to obtain the result. $\qquad\square$

We will derive Theorem IV.5.7 from Theorem IV.5.9, stated and proved below. This stronger result finds an appropriate coordinate system in which the formula for $\langle Az, z \rangle$ contains the principal minor determinants explicitly. It is helpful to first consider the one- and two-dimensional cases. When $n = 1$, a single real number determines the Hermitian form $\langle Az, z \rangle$:

$$\langle Az, z \rangle = p_1 | z_1 |^2.$$

When $n = 2$, one complex and two real numbers determine A:

$$\langle Az, z \rangle = p_1 | z_1 |^2 + \bar{a} z_1 \bar{z}_2 + a \bar{z}_1 z_2 + s | z_2 |^2.$$

The relationship $p_2 = p_1 s - | a |^2$ holds. Suppose that $p_1 \neq 0$. Completing the square yields

$$\langle Az, z \rangle = p_1 | z_1 + c z_2 |^2 + (s - p_1 | c |^2) | z_2 |^2,$$

where $p_1 c = a$. The coefficient of $| z_2 |^2$ is $\frac{p_2}{p_1}$. This statement follows by direct calculation:

$$s - p_1 |c|^2 = s - p_1 \frac{|a|^2}{p_1^2} = \frac{sp_1 - |a|^2}{p_1} = \frac{p_2}{p_1}.$$

One can also see it by using elementary linear algebra.

The proof of Theorem IV.5.9 combines this technique of completing the square with an induction argument.

Theorem IV.5.9. Suppose that A is Hermitian and that its leading principal minor determinants p_j for $1 \le j \le n$ do not vanish. Set $p_0 = 1$. Then there are complex numbers μ_j^k for $1 \le j \le n$ and $j+1 \le k \le n$, such that

$$\langle Az, z \rangle = \sum_{j=1}^{n} \frac{p_j}{p_{j-1}} \left| z_j + \sum_{k=j+1}^{n} \mu_j^k z_k \right|^2.$$

Proof. The result holds in the one-dimensional case, because

$$\langle Az, z \rangle = p_1 |z_1|^2 = \frac{p_1}{p_0} |z_1|^2.$$

Therefore we will assume that $n \ge 2$, and we write $z = (z_1, \dots, z_n)$.

Let r be an integer with $1 \le r \le n+1$. When $r \le n$ we let z' denote $(z_r, z_{r+1}, \dots, z_n)$. When $r = n+1$ we interpret z' as zero. For each r we will write

$$\langle Az, z \rangle = \sum_{j=1}^{r-1} \frac{p_j}{p_{j-1}} \left| z_j + \sum_{k=j+1}^{n} \mu_j^k z_k \right|^2 + \langle Bz', z' \rangle \qquad (13)$$

where $\langle Bz', z' \rangle$ is a Hermitian form. The proof of (13) uses induction on r. The basis step is simple. When $r = 1$ the summation on j from the right-hand side of (13) contains no terms, and thus the result holds with $B = A$.

Suppose we have established (13) for some r with $1 \le r \le n$. We will verify (13) with r replaced by $r+1$. By induction we then obtain (13) when $r = n+1$ and therefore the conclusion of the Theorem.

To help perform the induction step, we first write

$$\langle Bz', z' \rangle = B_{rr} |z_r|^2 + 2\text{Re} \sum_{k=r+1}^{n} B_{rk} z_r \bar{z}_k + \langle Bz'', z'' \rangle \qquad (14)$$

where $z'' = (z_{r+1}, \ldots, z_n)$. Set $z_l = 0$ for $l \geq r + 1$ in (14). By substituting the result in (13) we obtain an expression for $\langle Az, z \rangle$, assuming $z_l = 0$ for $l \geq r + 1$. Equating coefficients then yields (15) and (16):

$$A_{rr} = \sum_{l=1}^{r-1} \frac{p_l}{p_{l-1}} |\mu_l^r|^2 + B_{rr}. \qquad (15)$$

For each l with $1 \leq l < r$,

$$A_{rl} = \mu_l^r \frac{p_l}{p_{l-1}}. \qquad (16)$$

Solving (16) for μ_l^r, and plugging the result into (15) gives a formula for B_{rr}:

$$B_{rr} = A_{rr} - \sum_{l=1}^{r-1} \frac{p_{l-1}}{p_l} |A_{rl}|^2. \qquad (17)$$

By Lemma IV.5.8, (17) yields $B_{rr} = \frac{p_r}{p_{r-1}}$; hence also $B_{rr} \neq 0$.

Knowing that $B_{rr} \neq 0$ enables us to complete the square in (14). Completing the square gives, for some Hermitian C,

$$\langle Bz', z' \rangle = B_{rr} \left| z_r + \sum_{k=r+1}^{n} \mu_r^k z_k \right|^2 + \langle Cz'', z'' \rangle.$$

Combining (13) with the formula for $\langle Bz', z' \rangle$ yields

$$\langle Az, z \rangle = \sum_{j=1}^{r-1} \frac{p_j}{p_{j-1}} \left| z_j + \sum_{k=j+1}^{n} \mu_j^k z_k \right|^2$$

$$+ B_{rr} \left| z_r + \sum_{k=r+1}^{n} \mu_j^k z_k \right|^2 + \langle Cz'', z'' \rangle. \qquad (18)$$

Since $B_{rr} = \frac{p_r}{p_{r-1}}$, formula (18) gives (13) with r replaced by $r + 1$. By induction (13) holds for each r with $1 \leq r \leq n + 1$. When $r = n + 1$ we obtain the desired result.   □

For convenience we restate Theorem IV.5.9 in the following way.

Corollary IV.5.10. Suppose that A is Hermitian and that its leading principal minor determinants p_j for $1 \leq j \leq n$ do not vanish. Set $p_0 = 1$. Then there is a coordinate change $z \rightarrow \zeta(z)$ such that

$$\zeta_j(z) = z_j + \cdots$$

where the omitted terms denote a linear function in the variables z_k for $k > j$, and such that

$$\langle Az, z \rangle = \sum_{j=1}^{n} \frac{p_j}{p_{j-1}} |\zeta_j(z)|^2.$$

Proof. The corollary restates Theorem IV.5.9, so there is nothing to prove.   □

We finally prove Theorem IV.5.7.

Proof of Theorem IV.5.7. We first suppose that each leading principal minor determinant p_j is positive. Corollary IV.5.10 exhibits $\langle Az, z \rangle$ as a positive linear combination of squared absolute values of independent linear functions ζ_j. The positivity guarantees that A is nonnegative definite; the independence of the ζ_j guarantees that A is actually *positive* definite; see the proof of Theorem IV.5.2.

Next suppose that A is positive definite. For each integer k with $1 \leq k \leq n$ we let A' denote the corresponding leading principal submatrix. Each A' is positive definite, because $\langle A'z', z' \rangle = \langle Az, z \rangle$, where $z = (z', 0)$, and hence A' is invertible. Therefore $p_k = \det(A') \neq 0$, and Corollary IV.5.10 applies. The ζ_j must be independent linear functionals of z. We show that each p_k is positive by choosing appropriate values for ζ. We obtain $p_1 > 0$ by setting $\zeta_1 = 1$ and $\zeta_j(z) = 0$ for $j \geq 2$ in the formula from Corollary IV.5.10. By setting $\zeta_2(z) = 1$ and $\zeta_j(z) = 0$ otherwise, we obtain $\frac{p_2}{p_1} > 0$, and hence $p_2 > 0$. Continuing in this way we see that all the minor determinants are positive. $\square$

Suppose A is positive definite; it is easy to prove, without using Corollary IV.5.10, that the leading principal minor determinants are positive. Since A is positive definite, each leading principal submatrix A' is also positive definite. Corollary IV.5.1 guarantees that each eigenvalue of A' is positive. By elementary properties of determinants, the determinant of A' is the product of its eigenvalues and hence is a positive number.

Remark IV.5.11. Theorem IV.5.9 provides a nice interpretation of the leading principal minor determinants. We try to write the form $\langle Az, z \rangle$ as a sum of squared absolute values, by naively completing the square. The positivity of p_1 lets us get started, and allows us to eliminate z_1. The positivity of the r-th leading minor determinant enables us to get a positive coefficient in front of the term that eliminates z_r. Thus the positivity of the minor determinants is the condition needed for writing the form as a sum of squared absolute values.

Remark IV.5.12. The case where minor determinants can vanish requires care. Consider for example the matrix

$$A = \begin{pmatrix} 1 & 0 & 0 \\ 0 & 0 & 0 \\ 0 & 0 & -1 \end{pmatrix}.$$

Then A is Hermitian, and its leading principal minor determinants are $1, 0$, and 0. Yet A is obviously *not* nonnegative definite. There is an analogue of Theorem IV.5.7 for the degenerate case, but one must consider *all* principal minor determinants rather than only the *leading* principal minor determinants. See Exercise 16.

Suppose next that A is invertible, but has eigenvalues of both signs. We cannot conclude that all the leading principal minor determinants are nonzero. The simple example

$$A = \begin{pmatrix} 0 & 1 \\ 1 & 0 \end{pmatrix}$$

has $p_1 = 0$, but A is invertible. The correct statement is that there is a basis for which all the principal minor determinants are nonzero.

We pause here to mention the subject of Chapters VI and VII, namely, nonnegative polynomials on $\mathbf{C}^n$ of degree higher than two. Given a polynomial $p(z, \bar{z})$ with nonnegative values, it is natural to attempt to generalize the ideas from Theorem IV.5.7. Must p be a sum of squared absolute values (squared norm) of holomorphic polynomials? The answer is no in general. We will therefore try to write p as a *quotient* of squared norms. We will discover that nonnegative polynomials need not be quotients of squared norms. Upper triangular coordinate changes will also make an appearance in Chapter VI.

We close this section by deriving from our work several famous results about positive definite matrices. The first two results are inequalities due to Hadamard. Then we prove a curious result of Schur. Finally we include additional inequalities about positive definite matrices in the exercises.

Proposition IV.5.13. (Hadamard's inequality 1) Let A be a positive definite n-by-n matrix of complex numbers. Then

$$\det(A) \leq \prod_{j=1}^{n} A_{jj}.$$

Proof. By Theorem IV.5.7 each leading principal minor determinant p_j is positive. By Lemma IV.5.8 we then have

$$A_{nn} \geq \frac{p_n}{p_{n-1}}.$$

We apply the same reasoning to the principal submatrices of A and obtain, for $1 \leq k \leq n$,

$$A_{kk} \geq \frac{p_k}{p_{k-1}}.$$

Multiplying these inequalities together gives the conclusion. □

Corollary IV.5.14. (Hadamard's inequality 2) Let B be a square matrix of complex numbers, with column vectors b_j. Then

$$| \det(B) |^2 \leq \prod_{j=1}^{n} \| b_j \|^2.$$

Proof. When $\det(B) = 0$ the result is immediate. We may therefore assume that $\det(B) \neq 0$ and hence that B is invertible. Observe that $\det(B^*) = \overline{\det(B)}$. Let $A = B^*B$. Then A is positive definite by Theorem IV.5.2; hence

$$| \det(B) |^2 = \det(A) \leq \prod_{j=1}^{n} A_{jj} = \prod_{j=1}^{n} \| b_j \|^2. \qquad \square$$

We next briefly discuss a useful albeit curious result. The Schur product of two matrices of the same size is obtained by multiplying the entries together. We write $A \ s \ B$ for the matrix C for which $C_{jk} = A_{jk}B_{jk}$. In the proof we also use tensor product notation. Given vector spaces V and W of finite dimensions m and n and with bases $\{a_j\}$ and $\{b_j\}$, we may define a new vector space $V \otimes W$ of dimension mn with basis given by the mn elements written $a_j \otimes b_k$. Suppose V and W are equipped with inner products. The definition $\langle a \otimes b, a' \otimes b' \rangle = \langle a, a' \rangle \langle b, b' \rangle$ (and extension by linearity) provides an inner product on $V \otimes W$. Note that $\| v \otimes w \| = \| v \| \| w \|$, for $v \in V$ and $w \in W$.

Proposition IV.5.15. The Schur product of positive definite matrices is positive definite.

Proof. We sketch a proof based on Theorem IV.5.2. Since A and B are positive definite, we may assume that there are bases $\{a_j\}$ and $\{b_j\}$

such that $A_{kj} = \langle a_j, a_k \rangle$ and $B_{kj} = \langle b_j, b_k \rangle$. It then follows that

$$\langle (A \ s \ B)z, z \rangle = \sum_{j,k} A_{kj} B_{kj} z_j \bar{z}_k = \sum_{j,k} \langle a_j, a_k \rangle \langle b_j, b_k \rangle z_j \bar{z}_k$$

$$= \sum_{j,k} \langle a_j \otimes b_j, a_k \otimes b_k \rangle z_j \bar{z}_k$$

$$= \sum_{j} \| z_j (a_j \otimes b_j) \|^2 \geq 0.$$

It is also evident that we get strict inequality unless $z = 0$.   □

We offer an amusing corollary of this result. Suppose that $q : \mathbf{C} \to \mathbf{C}$ is a function. Define $q[[A]]$ to be the matrix with $q[[A]]_{kj} = q(A_{kj})$. Thus we apply the function q to each entry of A. When $q(x) = x^2$ we have $A \ s \ A = q[[A]]$.

Corollary IV.5.16. Let A be a positive definite matrix, and let q be a holomorphic function on $\mathbf{C}$ whose power series at the origin has all positive coefficients. Then $q[[A]]$ is positive definite.

Proof. By Proposition IV.5.15 and induction it follows that $q[[A]]$ is positive definite when $q(x) = x^n$ for some natural number n. A positive multiple of a positive definite matrix is itself positive definite. The sum of positive definite matrices is positive definite. Therefore $p[[A]]$ is positive definite when p is any partial sum of q. Note finally that the minimum eigenvalue of the sum of positive definite matrices exceeds the minimum eigenvalue of each summand. Therefore the minimum eigenvalue of the limiting matrix $q[[A]]$ is positive, and therefore $q[[A]]$ is positive definite.   □

It seems more natural to consider functions of the operator A, rather than functions applied entry by entry. Suppose that $q(z) = \sum a_n z^n$ is a convergent power series. We define $q(A)$ as expected by $q(A) = \sum a_n A^n$. A result corresponding to Corollary IV.5.16 holds for

$q(A)$. We leave the proof to Exercise 15. We close the section with additional exercises providing inequalities for positive definite matrices.

Exercise 15. Let A be positive definite, and let q be a holomorphic function on $\mathbf{C}$ whose power series at the origin has all positive coefficients. Prove that $q(A)$ is positive definite.

Exercise 16. Formulate a precise definition of (not necessarily leading) principal minor determinant. Suppose that *every* principal minor determinant of a Hermitian matrix A is nonnegative. Prove that A must be nonnegative definite. (See Remark IV.5.12.)

Exercise 17. Let A be positive definite. What is the minimum possible eigenvalue of $A + A^{-1}$?

Exercise 18. Suppose that A is positive definite. Show that

$$\det(\mathrm{Re}(A)) > \det(\mathrm{Im}(A)).$$

Exercise 19. Let A and B be matrices of the same size, and suppose $0 \le t \le 1$. We call the matrix $tA + (1-t)B$ a *convex combination* of A and B. Suppose that A and B are positive definite. Prove the following inequality:

$$\det(tA + (1-t)B) \ge (\det(A))^t (\det(B))^{1-t}.$$

The notation $C \le D$ means $D - C$ is nonnegative definite. Assume A and B are positive definite. Prove

$$(tA + (1-t)B)^{-1} \le tA^{-1} + (1-t)B^{-1}.$$

Exercise 20. (Minkowski) Let A and B be positive definite n-by-n matrices. Prove the following inequality:

$$(\det(A+B))^{\frac{1}{n}} \ge (\det(A))^{\frac{1}{n}} + (\det(B))^{\frac{1}{n}}.$$

IV.6   Hilbert's inequality

In the next two sections we apply Fourier analysis to obtain additional results on positive definiteness. The goal of this section is a famous inequality due to Hilbert:

$$\sum_{j,k=0}^{\infty} \frac{z_j \bar{z}_k}{1+j+k} \le \pi \sum_{k=0}^{\infty} |z_k|^2. \tag{19}$$

The matrix H whose j,k entry (for $j,k \ge 0$) is $\frac{1}{1+j+k}$ is called the *Hilbert* matrix. Hilbert's inequality says that the operator $\pi I - H$ is nonnegative definite on l^2. Note that the j,k entry of H depends only on the sum $j+k$. Matrices with this property are sometimes called *Hankel* matrices. We will also consider matrices whose j,k entry depends only on the difference $j-k$. These matrices are sometimes called *Toeplitz* matrices.

We derive Hilbert's inequality by way of Fourier series. Thus we will consider functions on the unit circle S^1. Note that a function f on $[0, 2\pi]$ such that $f(0) = f(2\pi)$ determines a function (also written f) on S^1, and conversely we may regard a function f on S^1 as a function on $[0, 2\pi]$ with $f(0) = f(2\pi)$. The basic question is to what extent f can be represented by a trigonometric series.

Fourier analysis offers powerful techniques for proving inequalities and deep insights into positivity conditions. Some of this material appears in Section IV.7; it is helpful but not strictly necessary for the main results in Chapters VI and VII.

Definition IV.6.1. (Trigonometric polynomials and trigonometric series) A *trigonometric polynomial* $p(t)$ is an expression

$$\sum_{k=-n}^{n} c_k e^{ikt}$$

where the coefficients c_k are complex numbers. The *degree* of p is n if $c_k = 0$ for $|k| > n$ and at least one of c_n and c_{-n} does not vanish. A *trigonometric series* is a formal sum

$$\sum_{k=-\infty}^{\infty} c_k e^{ikt}.$$

Observe that a trigonometric polynomial p determines a continuous complex-valued function on S^1. We may also regard p as a periodic function on **R** whose values are determined by its values on the interval $[0, 2\pi)$. Given a trigonometric polynomial p, thought of as a function on $[0, 2\pi]$, we can recover the coefficient c_k by taking the inner product with e^{ikt}:

$$c_k = \frac{1}{2\pi} \int_0^{2\pi} p(t) e^{-ikt}\, dt.$$

Now let f be an integrable function on $[0, 2\pi]$. We define its Fourier coefficients $c_k(f)$ as we did for trigonometric polynomials. Thus:

$$c_k(f) = \frac{1}{2\pi} \int_0^{2\pi} f(t) e^{-ikt}\, dt.$$

The formal trigonometric series S given by $\sum_k c_k(f) e^{ikt}$ is called the *Fourier series* of f. One of the basic questions of Fourier analysis is whether (and in what sense) the formal series $\sum_k c_k(f) e^{ikt}$ converges to f. We will only glimpse this beautiful and subtle subject.

To get started we compute the Fourier coefficients in a simple case; we then use this computation in proving Hilbert's inequality.

Example IV.6.2. Let a and b be complex numbers and put $g(t) = (a + bt)e^{-it}$. For $n \neq -1$ we have

$$c_n(g) = \frac{ib}{1+n}.$$

We also have $c_{-1} = a + b\pi$.

Proof. The elementary computation for $n = -1$ is left to the reader. It is clear for $n \neq -1$ that

$$\frac{1}{2\pi} \int_0^{2\pi} e^{-int} e^{-it} \, dt = \frac{1}{2\pi} \int_0^{2\pi} e^{-i(n+1)t} \, dt = 0.$$

Still assuming that $n \neq -1$ we compute

$$I_n = \frac{1}{2\pi} \int_0^{2\pi} e^{-int} t e^{-it} \, dt$$

by elementary integration by parts to obtain

$$I_n = \frac{1}{2\pi} \frac{t e^{-it(n+1)}}{-i(n+1)} \Big|_0^{2\pi} - \frac{1}{2\pi} \int_0^{2\pi} \frac{e^{-it(n+1)}}{-i(n+1)} \, dt = \frac{i}{1+n} - 0 = \frac{i}{1+n}.$$

The desired result follows from these two computations. $\square$

Theorem IV.6.3. (Hilbert's inequality) Suppose that $\{z_n\}$ (for $n \geq 0$) is a sequence of complex numbers and $\sum |z_n|^2$ is finite. Then

$$\sum_{j,k=0}^{\infty} \frac{z_j \overline{z}_k}{1+j+k} \leq \pi \sum_{k=0}^{\infty} |z_k|^2.$$

Proof. It follows from Example IV.6.2 that $\frac{1}{1+j+k}$ is the $(j+k)$-th Fourier coefficient of g, where $g(t) = i(\pi - t)e^{-it}$. Below we will explain why we choose the constant $i\pi$. For each N, we consider the matrix with entries $\frac{1}{1+j+k}$ for $0 \leq j, k \leq N$. Writing $\frac{1}{1+j+k}$ as an integral yields

$$\sum_{j,k=0}^{N} \frac{z_j \overline{z}_k}{1+j+k} = \frac{1}{2\pi} \int_0^{2\pi} \sum_{j,k=0}^{N} z_j \overline{z}_k e^{-i(j+k)t} g(t) \, dt$$

$$= \frac{1}{2\pi} \int_0^{2\pi} \left(\sum_{j=0}^{N} z_j e^{-ijt} \right) \left(\sum_{k=0}^{N} \overline{z}_k e^{-ikt} \right) g(t) \, dt$$

$$\leq \frac{1}{2\pi} \int_0^{2\pi} |g(t)| \, \Big| \sum_{j=0}^{N} z_j e^{-ijt} \Big| \, \Big| \sum_{k=0}^{N} \overline{z}_k e^{-ikt} \Big| \, dt.$$

$$(20)$$

Note that $|g(t)| = |\pi - t|$. We will be replacing $|g(t)|$ by its maximum, so we have chosen (in the notation of Example IV.6.2) $a = i\pi$ to minimize the maximum of $|g(t)|$ on $[0, 2\pi]$. This maximum equals π. After replacing $|g(t)|$ by π, we will estimate the integral by using the Cauchy-Schwarz inequality. Since the distinct exponentials are orthogonal, we have

$$\int_0^{2\pi} \left| \sum_{j=0}^{N} z_j e^{-ijt} \right|^2 dt = \int_0^{2\pi} \left| \sum_{k=0}^{N} \overline{z}_k e^{-ikt} \right|^2 dt = (2\pi) \sum_{j=0}^{N} |z_j|^2.$$

(21)

In the application of the Cauchy-Schwarz inequality, each of the two sums from the last term in (20) thus contributes the square root of the right-hand side in (21). Thus (20) and (21) yield

$$\sum_{j,k=0}^{N} \frac{z_j \overline{z}_k}{1 + j + k} \leq \pi \sum_{n=0}^{N} |z_n|^2. \qquad (22)$$

Since (22) holds for each N, we may let N tend to infinity and thus obtain Hilbert's inequality.   □

This proof of Hilbert's inequality yields the following more general statement.

Theorem IV.6.4. Suppose that g is real-valued, integrable, and bounded on $[0, 2\pi]$ with $\sup(g) = M$. Let $c_k(g)$ be its k-th Fourier coefficient. Let $c_{jk} = c_{j+k}(g)$ define the corresponding Hankel matrix. The following inequality holds:

$$\sum_{j,k=0}^{\infty} c_{jk} z_j \overline{z}_k \leq M \sum_{j=0}^{\infty} |z_j|^2.$$

Proof. We repeat the proof of Hilbert's inequality starting with

$$\sum_{j,k=0}^{N} c_{jk} z_j \overline{z}_k = \frac{1}{2\pi} \int_0^{2\pi} \sum_{j,k=0}^{N} z_j \overline{z}_k e^{-i(j+k)t} g(t)\, dt$$

and hence obtain

$$\sum_{j,k=0}^{N} c_{jk} z_j \bar{z}_k \leq \frac{M}{2\pi} \int_0^{2\pi} \left| \sum_{j,k=0}^{N} z_j \bar{z}_k e^{-i(j+k)t} \right| dt$$

$$\leq \frac{M}{2\pi} \int_0^{2\pi} \left| \sum_{j=0}^{N} z_j e^{-ijt} \right| \left| \sum_{k=0}^{N} \bar{z}_k e^{-ikt} \right| dt$$

$$\leq M \sum_{j=0}^{N} |z_j|^2. \tag{23}$$

We clarify the last step. The sums differ because of complex conjugation. We estimate by the Cauchy-Schwarz inequality and use (21) as before. We obtain, for each N,

$$\sum_{j,k=0}^{N} c_{jk} z_j \bar{z}_k \leq M \sum_{j=0}^{N} |z_j|^2.$$

The theorem follows by letting N tend to infinity. $\square$

Notice how the proof of Theorem IV.6.4 would be nicer if we considered Toeplitz matrices c_{j-k} rather than Hankel matrices c_{j+k}. In the Toeplitz case we obtain

$$\sum_{j,k=0}^{\infty} c_{jk} z_j \bar{z}_k = \frac{1}{2\pi} \int_0^{2\pi} \sum_{j,k=0}^{\infty} z_j \bar{z}_k e^{-i(j-k)t} g(t)\, dt$$

$$= \frac{1}{2\pi} \int_0^{2\pi} |\sum_{j=0}^{\infty} z_j e^{-ijt}|^2 g(t)\, dt. \tag{24}$$

Notice that the Cauchy-Schwarz inequality is not used in (24).

Formula (24) becomes particularly useful when we have bounds on g. For example, when g is nonnegative, the Toeplitz matrix formed from the Fourier coefficients of g is necessarily nonnegative definite. One can also replace the function g by a positive measure. We investigate these ideas in the next section.

IV.7 Additional inequalities from Fourier analysis

The proofs of Hilbert's inequality and its generalization Theorem IV.6.4 indicate the power of Fourier analysis in proving and understanding inequalities. In this section we establish enough information about Fourier series to prove several additional results.

We prove Fejér's theorem that the Cesàro means $\sigma_N(f)$ of the Fourier series of a continuous function f on S^1 converge uniformly to f. Thus trigonometric polynomials are dense in $C(S^1)$. Density of the trigonometric polynomials easily implies (see Exercise 29) the Weierstrass approximation theorem, which states that ordinary polynomials are dense in $C([a, b])$.

Motivated by our interest in positivity conditions, we prove Herglotz's theorem characterizing nonnegative Toeplitz matrices as those coming from the Fourier coefficients of a positive measure. After defining the Fourier transform on Euclidean space $\mathbf{R}^n$, we state without proof Bochner's generalization of the Herglotz theorem to functions of positive type. The section closes with Wirtinger's inequality, which will be revisited in Chapter V.

Let us first make some remarks about measures and then give the definition of Fourier coefficients of a measure. The easiest way to introduce measures is as continuous linear functionals. We therefore first recall the definition of the dual space of a Banach space.

Definition IV.7.1. Suppose that V is a (complex) Banach space. Its dual space V^* is the space of continuous linear functionals on V.

The Riesz Lemma (Theorem II.2.4) shows that the dual space of a Hilbert space may be identified with the space itself. The Riesz Lemma is a special case of more general Riesz *Representation Theorems*. See [F]. Such theorems characterize the dual spaces of various Banach spaces. Let X be a compact subset of $\mathbf{R}^k$, and let $C(X)$ be the space of continuous complex-valued functions on X, as discussed after Definition II.1.5. All we need to know here is that the dual space of $C(X)$ is the space of complex measures on X; if the reader is unfamiliar with measure theory, then this statement may be used as the *definition* of complex measure.

We clarify the notion of measure using informal language. Suppose L assigns a number $L(f)$ to each continuous function f on X. If L is linear, and there is a constant C such that $|L(f)| \leq C||f||_\infty$ for all f, then there is a measure μ such that $L(f) = \int_X f \, d\mu$. The measure μ is called *positive* if $\int_X f \, d\mu \geq 0$ whenever $f \geq 0$. Although this usage of *positive* is standard, it might be preferable to say *nonnegative* in this context.

We mention a simple point about dual operators. Let $K : V \to V$ be a bounded linear transformation. There is a corresponding *dual* linear transformation $K^* : V^* \to V^*$ defined by

$$K^*(L)(f) = L(K(f)). \tag{25}$$

It is easy to see that K^* is bounded and that $||K^*|| = ||K||$.

We now define the Fourier coefficients of a measure.

Definition IV.7.2. Let g be a finite measure on $[0, 2\pi]$. The Fourier coefficients $c_k(g)$ of g are defined by

$$c_k(g) = \frac{1}{2\pi} \int_0^{2\pi} e^{-ikt} \, dg(t). \tag{26}$$

Example IV.7.3. The *Dirac measure* δ is the linear functional on $C([0, 2\pi])$ defined by $\delta(f) = f(0)$. It is immediate from (26) that $c_n(\delta) = 1$ for all n.

In case g is a positive measure, and $c_{jk} = c_{j-k}(g)$, (24) shows that

$$\sum_{j,k=0}^{\infty} c_{jk} z_j \overline{z}_k = \frac{1}{2\pi} \int_0^{2\pi} \left| \sum_{j=0}^{\infty} z_j e^{-ijt} \right|^2 dg(t) \geq 0.$$

This inequality provides the proof of the easy direction of Theorem IV.7.10 below. Herglotz characterized nonnegative definite Toeplitz matrices by establishing the converse assertion.

Let $S = \sum a_n e^{int}$ be a trigonometric series. Its symmetric partial sums are given by

$$s_n(S, t) = \sum_{|j| \leq n} a_j e^{ijt}.$$

Each $s_n(S, t)$ is a continuous function of t; if the symmetric partial sums converge uniformly on $[0, 2\pi]$, then the limit function f must be continuous. It follows that S is the Fourier series for f. A fundamental aspect of Fourier analysis is that the Fourier series of a continuous function does not in general converge uniformly. In fact the Fourier series of a continuous function can diverge at large sets of points. See [Ka].

Because it is difficult to analyze convergence of the partial sums, one considers instead (compare Exercise 10 from Chapter I) the Cesàro means.

Definition IV.7.4. Let $S = \sum a_n e^{int}$ be a trigonometric series. Its *Cesàro mean* of order N, written $\sigma_N(S, t)$, is defined by

$$\sigma_N(S, t) = \sum_{|j| \leq N} \left(1 - \frac{|j|}{N + 1} \right) a_j e^{ijt}.$$

The Cesàro mean is the average of the symmetric partial sums (Exercise 28). Both s_n and σ_N define bounded operators on $C(S^1)$ (Exercise 30). It is customary to let s_n and σ_N also denote the corresponding dual operators. These would have stars if we followed the notation in (25).

The next result, which is a simple version of Parseval's formula, will get used in the proof of Herglotz's theorem.

Proposition IV.7.5. (Parseval's formula) Let P be a trigonometric polynomial, and let $S = \sum a_j e^{ijt}$ be a trigonometric series. Then

$$\sum c_n(P)\overline{a}_n = \lim_{N \to \infty} \frac{1}{2\pi} \int_0^{2\pi} P(t)\overline{\sigma_N(S, t)} \, dt. \qquad (27)$$

Proof. Suppose P is of degree D. We have

$$\frac{1}{2\pi}\int_0^{2\pi} P(t)\overline{\sigma_N(S,t)}\,dt = \frac{1}{2\pi}\int_0^{2\pi}\sum_{|n|\leq D}\sum_{|j|\leq N} c_n(P)e^{int}$$

$$\times\left(1-\frac{|j|}{N+1}\right)\overline{a}_j e^{-ijt}\,dt.$$

By the orthogonality of the exponentials $\{e^{ikt}\}$ and under the assumption that $N \geq D$, the right-hand side becomes

$$\sum_{|n|\leq D}(1-\frac{|n|}{N+1})c_n(P)\overline{a}_n.$$

Letting $N \to \infty$ gives the desired conclusion. $\qquad\square$

In case both P and S are, for example, trigonometric polynomials, formula (27) has a simple interpretation. The left-hand side of (27) is the inner product in l^2 of the sequences of Fourier coefficients; the right-hand side is the inner product $\langle P, S\rangle$ in $L^2(S^1)$ of the functions. Thus the *Fourier transform* is an isometry.

The Cesàro means provide, via Fejér's Theorem (IV.7.7), a proof that trigonometric polynomials are dense in $C(S^1)$. This fact has many applications; in particular it will help us prove Herglotz's theorem.

Consider the trigonometric series S defined by $\sum e^{int}$; by Example IV.7.3 it arises from the Dirac measure δ. Its Cesàro means $\sigma_N(S,t)$ then serve as approximations to δ. We demonstrate this in the next lemma. Formula (29) there provides an elementary expression for the Cesàro means $\sigma_N(S,t)$ which makes their positivity evident. Formula (28) shows that $\frac{\sigma_N(S,t)\,dt}{2\pi}$ is a *probability measure* on the circle. Formula (30) indicates that the probability becomes more concentrated about a single point as N tends to infinity. We thus realize the *Dirac delta function* as a limit of functions.

Lemma IV.7.6. (Approximation to the Dirac measure) Let S denote the trigonometric series $\sum e^{int}$. The Cesàro means $\sigma_N(S,t)$ satisfy the following three formulas:

$$\frac{1}{2\pi} \int_0^{2\pi} \sigma_N(S, t)\, dt = 1,$$ (28)

$$\sigma_N(S, t) = \frac{1}{N+1} \frac{\sin^2(\frac{N+1}{2}t)}{\sin^2(\frac{t}{2})},$$ (29)

and for each η with $0 < \eta < \pi$,

$$\lim_{N \to \infty} \frac{1}{2\pi} \int_\eta^{2\pi - \eta} \sigma_N(S, t)\, dt = 0.$$ (30)

Proof. Formula (28) is immediate. *Deriving* (29) is a beautiful exercise in manipulating the finite geometric series (Exercise 23). Verifying (29) is easy. We have

$$\sin^2\left(\frac{t}{2}\right) = -\frac{e^{-it}}{4} + \frac{1}{2} - \frac{e^{it}}{4}.$$

Multiplying the right-hand side of this formula by $\sigma_N(S, t)$ gives

$$\sin^2\left(\frac{t}{2}\right)\sigma_N(S, t) = \left(-\frac{e^{-it}}{4} + \frac{1}{2} - \frac{e^{it}}{4}\right) \sum_{|j| \le N} \left(1 - \frac{|j|}{N+1}\right) e^{ijt}$$

$$= \frac{1}{N+1}\left(-\frac{e^{-i(N+1)t}}{4} + \frac{1}{2} - \frac{e^{i(N+1)t}}{4}\right)$$

$$= \frac{1}{N+1} \sin^2\left(\frac{N+1}{2}t\right).$$

We have now verified (29).

We now use (29) to prove (30). Fix η with $0 < \eta < \pi$. The positive expression

$$\frac{\sin^2(\frac{N+1}{2}t)}{\sin^2(\frac{t}{2})}$$

is bounded above by some number M independently of N on the interval $[\eta, 2\pi - \eta]$, because we keep away from the zeroes of $\sin(\frac{t}{2})$. Hence

$$\frac{1}{2\pi} \int_\eta^{2\pi-\eta} \sigma_N(S, t)\, dt \le \frac{2\pi - 2\eta}{2\pi} \frac{M}{N+1}.$$

Letting $N \to \infty$ we obtain (30). $\square$

Exercise 21. Graph $\sigma_N(S, t)$ for $N = 1, 2, 3, 4$.

From (29) we see that $\sigma_N(S, t)$ is nonnegative. Properties (28) and (30) then guarantee that the Cesàro means provide an approximation to the Dirac measure. This result enables us to prove Fejér's Theorem. In the proof we assume that the domain of f has been extended past $[0, 2\pi]$ so that f is periodic.

Theorem IV.7.7. (Fejér) Let f be continuous on S^1. Then the Cesàro means $\sigma_N(f)$ of its Fourier series converge uniformly to f on S^1.

Proof. Choose $x \in [0, 2\pi]$. We will explicitly estimate $|f(x) - \sigma_N(f, x)|$ and show that it tends to zero uniformly in x. Thus given $\epsilon > 0$, we want to find an N_0 such that, for all x, and for $N \ge N_0$,

$$|f(x) - \sigma_N(f, x)| < \epsilon.$$

By the definition of $\sigma_N(f)$ we have

$$f(x) - \sigma_N(f, x)$$
$$= \frac{1}{2\pi} \int_0^{2\pi} \left(f(x) - \sum_{|j| \le N} \left(1 - \frac{|j|}{N+1} \right) e^{-ij(t-x)} f(t) \right) dt.$$

We translate the integrand by writing $t - x = u$. We use the periodicity of f, and we write the limits of integration from $-\pi$ to π for convenience. Then the definition of $\sigma_N(S, u)$ and (28) give

$$f(x) - \sigma_N(f, x) = \frac{1}{2\pi} \int_{-\pi}^{\pi} (f(x) - f(x + u))\sigma_N(S, u) \, du$$

and hence

$$| f(x) - \sigma_N(f, x) | \le \frac{1}{2\pi} \int_{-\pi}^{\pi} | f(x) - f(x + u) | \sigma_N(S, u) \, du. \quad (31)$$

We are given a positive ϵ. We use it to decide where to break the interval of integration in (31) into two pieces. Since f is continuous on the circle, and the circle is a compact set, f is *uniformly continuous* there. By the uniform continuity of f there is an $\eta > 0$ such that $| f(x) - f(x + u) |$ is uniformly small in the interval where $| u | \le \eta$. Since the integral over this interval of $\sigma_N(S, u)$ is bounded by unity, we can choose η to ensure that the contribution to (31) for $| u | \le \eta$ is at most $\frac{\epsilon}{2}$. Outside of this interval, we bound $| f(x) - f(x + u) |$ by a constant. Then (30) shows that we can choose N sufficiently large such that the integral over the second piece is also at most $\frac{\epsilon}{2}$. Hence $\sigma_N(f)$ tends to f uniformly. $\qquad \square$

Corollary IV.7.8. Trigonometric polynomials are dense in $C(S^1)$. Thus, given $f \in C(S^1)$ and $\epsilon > 0$, there is a trigonometric polynomial P such that $\| f - P \|_\infty < \epsilon$.

Proof. Given $f \in C(S^1)$, the Cesàro means $\sigma_N(f)$ of its Fourier series are trigonometric polynomials, and they converge uniformly to f by Fejér's theorem. In other words, given $\epsilon > 0$, we can find an N such that $\| f - \sigma_N(f) \|_\infty < \epsilon$. $\qquad \square$

Remark IV.7.9. Suppose that f is continuously differentiable on $[0, 2\pi]$ and $f(0) = f(2\pi)$. Using integration by parts one can show that $s_n(f, t)$ converges uniformly (and hence pointwise) to f; in this case we may write

$$f(t) = \sum_{k=-\infty}^{\infty} c_k(f)e^{ikt}.$$

This equality holds for all t, and provides a simple circumstance under which the Fourier series precisely represents the function.

We are ready to prove Herglotz's theorem characterizing the nonnegativity of Toeplitz matrices in terms of Fourier coefficients of positive measures.

Theorem IV.7.10. (Herglotz) Let b_n be a countable collection of complex numbers, indexed by the integers, such that $b_{-n} = \overline{b}_n$. The Toeplitz matrix (b_{j-k}) for $0 \leq j, k \leq N$ is nonnegative definite for each N if and only if there is a positive measure g on $[0, 2\pi]$ with $b_n = c_n(g)$ for all n.

Proof. Suppose first that g is a positive measure on $[0, 2\pi]$. We put $b_n = c_n(g)$ and compute

$$\sum_{j,k=0}^{N} b_{j-k} z_j \overline{z}_k = \sum_{j,k=0}^{N} \frac{1}{2\pi} \int_0^{2\pi} e^{-i(j-k)t} z_j \overline{z}_k \, dg(t)$$

$$= \frac{1}{2\pi} \int_0^{2\pi} \left| \sum_{j=0}^{N} z_j e^{-ijt} \right|^2 dg(t) \geq 0.$$

Hence the Toeplitz matrix b_{j-k} for $0 \leq j, k \leq N$ is nonnegative definite.

Conversely, suppose we are given the sequence $\{b_n\}$; we must construct the measure. Let B denote the trigonometric series $\sum b_j e^{ijt}$, and let $\sigma_N(B, t)$ denote its N-th Cesàro mean.

The Toeplitz matrix (b_{j-k}) is nonnegative; we choose $z_j = e^{ijt}$ and obtain for each N

$$0 \leq \sum_{j,k=0}^{N} b_{j-k} z_j \overline{z}_k = \sum_{j,k=0}^{N} b_{j-k} e^{-i(j-k)t} = (N+1)\sigma_N(B, t). \quad (32)$$

Hence each Cesàro mean is nonnegative. In Exercise 24 you are asked to check the equality on the far right in (32).

Consider the trigonometric polynomial $\sigma_N(B, t)$ as an element of the dual space of $C(S^1)$; that is, $\sigma_N(B, t)$ is a complex measure on the circle. We compute its norm $|| \sigma_N(B, t) ||_*$ as a linear functional. Because $\sigma_N(B, t)$ is nonnegative, the supremum in the definition of norm of a linear transformation is achieved with $f = 1$. Therefore

$$|| \sigma_N(B, t) ||_* = \sup_{(|| f ||_\infty = 1)} \frac{1}{2\pi} \int_0^{2\pi} f(t)\sigma_N(B, t)\, dt$$

$$= \frac{1}{2\pi} \int_0^{2\pi} \sigma_N(B, t)\, dt = b_0.$$

The positive linear functionals defined by the $\sigma_N(B, t)$ thus all have the same norm.

Let V denote the subspace of $C(S^1)$ consisting of trigonometric polynomials. It is dense in $C(S^1)$ by Corollary V.7.8. We define a linear functional g on V by giving the value of $g[P]$ for $p \in V$ as in the first equality in (33) below. Here $P = \sum_{-D}^D c_n(P)e^{int}$. Since $\sigma_N(B, t)$ is real, the second equality in (33) follows immediately from Proposition IV.7.5:

$$g[P] = \sum_{-D}^D c_n(P)\overline{b}_n = \lim_{N \to \infty} \frac{1}{2\pi} \int_0^{2\pi} P(t)\sigma_N(B, t)\, dt. \quad (33)$$

Since the norms of the linear functionals $\sigma_N(B, t)$ all equal b_0, the norm of g is at most b_0. Since V is dense in $C(S^1)$, we see that g extends to a bounded linear functional on $C(S^1)$, and hence defines a measure. Since each $\sigma_N(B, t)$ is a positive linear functional, so is g. $\qquad\square$

Bochner proved a beautiful generalization of Herglotz's theorem for functions of several variables. In order to state this generalization we need to give the definition of the Fourier transform on real Euclidean space. We will write $x \cdot y$ for the usual Euclidean inner product.

Definition IV.7.11. (Fourier transform on $\mathbf{R}^n$) Suppose that f is integrable on $\mathbf{R}^n$. We define its Fourier transform $f^\wedge$ as a function on $\mathbf{R}^n$

by

$$f^\wedge(\xi) = (2\pi)^{-\frac{n}{2}} \int_{\mathbf{R}^n} e^{-ix\cdot\xi} f(x)\, dV(x).$$

As in the case of Fourier series we can also take the Fourier transform of a finite measure. We refer to [Do], [F], [Ka], or [S] for detailed information.

For convenience in stating Bochner's result we introduce the following terminology.

Definition IV.7.12. Let $f : \mathbf{R}^n \to \mathbf{C}$ be a function. We say that f is of *positive type* if the following holds. For all positive integers N and all choices of N points $x_j \in \mathbf{R}^n$, the N-by-N matrix whose j, k entry is $f(x_j - x_k)$ is nonnegative definite.

Suppose that f is of positive type. Letting $N = 1$ shows that $f(0)$ is nonnegative. Recall from Proposition IV.1.4 that nonnegative definite matrices are Hermitian. When $N = 2$ we obtain $f(-x) = \overline{f(x)}$; then invoking the principal minors test (Theorem IV.5.7) shows that $|f(x)| \le f(0)$ for all x. Theorem IV.5.7 yields additional inequalities, but quickly leads to a mess. This mess contrasts with the beautiful theorem of Bochner, which characterizes continuous functions of positive type. See [Do] or [S] for a proof. The reader should also compare this result with Application IV.5.5 and its generalization.

Theorem IV.7.13. (Bochner) A continuous function on $\mathbf{R}^n$ is of positive type if and only if it is the Fourier transform of a finite positive measure.

Remark IV.7.14. It is not difficult to derive the one-dimensional case of Bochner's theorem from Herglotz's theorem. See [Ka] for example.

Remark IV.7.15. Bochner's theorem holds for locally compact Abelian groups. Both Fourier transforms and functions of positive type make sense in this context. The abstract general version (due to Weil) contains both Theorem IV.7.10 and Theorem IV.7.13 as special cases; in the first case the group is the integers and in the second case the group is real Euclidean space.

We close this section with an inequality relating the L^2 norm of a function on $[0, 2\pi]$ to the L^2 norm of its derivative. We will provide a different proof in Application V.3.8.

Application IV.7.16. (**Wirtinger's inequality**) Let f be continuously differentiable on an open interval containing $[0, 2\pi]$, with $f(0) = f(2\pi)$. We also suppose that the average value of f is zero, that is,

$$\int_0^{2\pi} f(t)\, dt = 0.$$

Let $\| f \|_2$ denote the usual L^2 norm of f on $[0, 2\pi]$. Then

$$\| f \|_2 \leq \| f' \|_2,$$

and equality occurs if and only if f is a linear combination of cosine and sine.

Proof. Because f is continuously differentiable, by Remark IV.7.9 it is given by its Fourier series. Suppose therefore that $f(t) = \sum_{-\infty}^{\infty} a_n e^{int}$. The average value condition guarantees that $a_0 = 0$. We have $\| f \|_2^2 = \sum |a_n|^2$ while $\| f' \|_2^2 = \sum n^2 |a_n|^2$. Therefore $\| f \|_2 \leq \| f' \|_2$, and equality can hold only if $a_n = 0$ for $|n| \neq 1$. Therefore equality holds only if

$$f(t) = a_1 e^{it} + a_{-1} e^{-it} = (a_1 + a_{-1}) \cos(t) + i(a_1 - a_{-1}) \sin(t),$$

as was claimed.   □

Application IV.7.17. (**A relative of Wirtinger's inequality**) Let us drop the condition on the average value of f in Application IV.7.16, but continue to assume that f is continuously differentiable in an interval containing $[0, 2\pi]$; in particular $f' \in L^2(S^1)$ and $f(t) = \sum_{-\infty}^{\infty} a_n e^{int}$. The following inequality holds on the Fourier coefficients a_n:

$$\sum_{-\infty}^{\infty} |a_n| \le \| f \|_1 + \frac{\pi}{\sqrt{3}} \| f' \|_2.$$

Proof. For $n \ne 0$ we write $a_n = \frac{1}{n} n a_n$ and use the Cauchy-Schwarz inequality in l^2. We obtain the inequality

$$\sum_{-\infty}^{\infty} |a_n| = |a_0| + \sum_{n \ne 0} |a_n| \le |a_0| + \left(\sum_{n \ne 0} \frac{1}{n^2} \right)^{\frac{1}{2}} \left(\sum_{n \ne 0} |n a_n|^2 \right)^{\frac{1}{2}}.$$

Next we observe that $|a_0| \le \| f \|_1$ and that $\sum_{n \ne 0} |n a_n|^2 = \| f' \|_2^2$. From Exercise 25 we may substitute $\frac{\pi^2}{3}$ for $\sum_{n \ne 0} \frac{1}{n^2}$ and complete the proof. $\qquad \square$

Remark IV.7.18. The Wirtinger inequalities provide control on a function in terms of control on its derivative. Inequalities relating norms of a function of several variables to norms of its gradient are known as Poincaré inequalities. These play significant roles in harmonic analysis and in partial differential equations.

Exercise 22. For $f \in L^1(S^1)$, verify that $|a_0| \le \| f \|_1$.

Exercise 23. Derive formula (29) using the finite geometric series.

Exercise 24. Show that $\sum_{j,k=0}^{N} b_{j-k} e^{-i(j-k)t} = (N+1)\sigma_N(B, t)$ (from the proof of Theorem IV.7.10).

Exercise 25. Prove that $\sum_{n=1}^{\infty} \frac{1}{n^2} = \frac{\pi^2}{6}$. Suggestion 1: Compute the Fourier series on $[0, 2\pi]$ for the function f defined there by $f(t) = (\pi - t)^2$. Since $f(0) = f(2\pi)$ and f is continuously differentiable, its Fourier series converges to $f(t)$ at each point t.

Suggestion 2: Evaluate the double integral

$$\int_0^1 \int_0^1 \frac{dx\,dy}{1-xy}.$$

First expand the integrand in a geometric series and integrate term by term to obtain $\sum_{n=1}^{\infty} \frac{1}{n^2}$. To evaluate the integral in closed form, make the change of variables $(x, y) = (u + v, u - v)$. Compute the resulting integral by breaking the region of integration into two pieces and then using basic calculus.

Exercise 26. (Open problem) Try to evaluate the integral

$$\int_0^1 \int_0^1 \int_0^1 \frac{dx\,dy\,dz}{1-xyz}$$

to obtain a value for $\sum_{n=1}^{\infty} \frac{1}{n^3}$.

Comment on Exercises 25 and 26. The Riemann zeta function is defined for $s > 1$ by the series

$$\zeta(s) = \sum_{n=1}^{\infty} \frac{1}{n^s}.$$

Exercise 24 provides two ways to find $\zeta(2)$ explicitly; methods of Fourier series work for evaluating ζ at even integers, but these methods break down at odd integers. The computation of integrals similar to those in Exercises 25 and 26 have led to recent results about $\zeta(3)$; see [Hu] for interesting information about these integrals, for additional references, and for a slick way to evaluate the integral in Exercise 25.

Exercise 27. Define p by $p(x) = 1 + a \cos(x) + b \cos(2x)$. Find the necessary and sufficient conditions on a and b so that $p(x) \geq 0$ for all x. Use calculus and also use Herglotz's theorem.

Exercise 28. (Cesàro means and partial sums) Show that the Cesàro means (Definition IV.7.4) are the averages of the symmetric partial sums.

Exercise 29. **(Weierstrass approximation)** Use Corollary IV.7.8 to prove that ordinary polynomials are dense in $C([a, b])$. Suggestion: First show that it suffices to assume $[a, b] = [0, 1]$. Then, for f continuous on $[0, 1]$, consider the function $t \rightarrow f(|\cos(t)|)$. Approximate it by trigonometric polynomials. By Exercise 14 from Chapter I each function $\cos(jt)$ is a polynomial in $\cos(t)$.

Exercise 30. Show that both s_n and σ_N define bounded operators on $C(S^1)$.

Exercise 31. Suppose that f is continuously differentiable with Fourier series S. Prove that $s_n(S, t)$ and $\sigma_n(S, t)$ have the same limit at each point. Then verify Remark IV.7.9.

CHAPTER V

Compact and Integral Operators

This chapter discusses compact operators in detail. Section 1 concerns convergence properties for sequences of bounded operators. We introduce compact operators in Section 2. The main result of the chapter is the spectral theorem for compact Hermitian operators. We provide additional discussion about integral operators, and we close the chapter with a glimpse at singular integral operators. This chapter prepares us for the applications in Chapter VII, where we use facts about compact operators to study positivity conditions for polynomials.

V.1 Convergence properties for bounded linear transformations

Completeness of the real number system is crucial for doing analysis; without it *limits* would be a useless concept. Similarly, completeness for metric spaces (such as Hilbert and Banach spaces) is needed for doing analysis in more general settings. We therefore begin this chapter with a short discussion of three possible notions for convergence of sequences of bounded linear transformations between Hilbert spaces.

We write $\mathcal{L}(\mathcal{H}, \mathcal{H}')$ for the vector space of bounded linear transformations between Hilbert spaces $\mathcal{H}$ and $\mathcal{H}'$. We mentioned in Section II.1 that $\mathcal{L}(\mathcal{H}, \mathcal{H}')$ is a Banach space; in particular this space is complete. See Theorem V.1.4.

We next consider three reasonable definitions of convergence in $\mathcal{L}(\mathcal{H}, \mathcal{H}')$.

Definition V.1.1. (Notions of convergence for bounded linear transformations) Let $\{L_n\}$ be a sequence in $\mathcal{L}(\mathcal{H}, \mathcal{H}')$, let $L \in \mathcal{L}(\mathcal{H}, \mathcal{H}')$.

1) $\{L_n\}$ converges to L *in norm* if the sequence of real numbers $\| L_n - L \|$ converges to 0.

2) $\{L_n\}$ converges to L *in the strong operator topology* if, for all $z \in \mathcal{H}$, the sequence $\{L_n(z)\}$ converges to $L(z)$ in $\mathcal{H}'$.

3) $\{L_n\}$ converges to L *in the weak operator topology* if, for all $z \in \mathcal{H}$ and all $w \in \mathcal{H}'$, the sequence of complex numbers $\langle L_n(z), w \rangle$ converges to $\langle L(z), w \rangle$.

Lemma V.1.2. If $\{L_n\}$ converges to L in the strong operator topology, then $\{L_n\}$ converges to L in the weak operator topology. If $\{L_n\}$ converges to L in norm, then it converges to L in the strong operator topology and hence in the weak operator topology.

Proof. By considering $L_n - L$ we reduce to the case where $\{L_n\}$ converges to the zero operator. Then 1) means that $\| L_n \|$ converges to 0, 2) means that $\| L_n(z) \|$ converges to 0 for all z, and 3) means that $| \langle L_n z, w \rangle |$ converges to 0 for all z and w.

By the Cauchy-Schwarz inequality $| \langle L_n z, w \rangle | \leq \| L_n(z) \| \| w \|$. Hence we see that 2) implies 3). By definition of the norm on $\mathcal{L}(\mathcal{H}, \mathcal{H}')$ we have $\| L_n(z) \| \leq \| L_n \| \| z \|$. Now we see that 1) implies 2). $\qquad \square$

The following simple example shows that reversing the implications in Lemma V.1.2 is not possible.

Example V.1.3. Let $\mathcal{H}$ be an infinite-dimensional Hilbert space with complete orthonormal system $\{\phi_j\}$. We consider three sequences of operators in $\mathcal{L}(\mathcal{H}, \mathcal{H})$:

Let L_n be defined by $L_n(z) = \frac{z}{n}$. Then $\{L_n\}$ converges to the zero operator in norm and hence also in the strong and weak operator topologies.

Let P_n be the operator defined by

$$P_n(z) = \sum_{j=n+1}^{\infty} \langle z, \phi_j \rangle \phi_{j-n}.$$

Since $P_n(\phi_{n+1}) = \phi_1$, the norm of P_n is at least one. Thus $\{P_n\}$ does not tend to zero in norm. It does converge to 0 in the strong operator topology. To see this, observe that $\| P_n(z) \|^2 = \sum_{n+1}^{\infty} | \langle z, \phi_j \rangle |^2$; this is the tail of a convergent sum for $\| z \|^2$ and hence tends to zero.

Let Q_n be the n-shift defined by

$$Q_n(z) = \sum_{j=1}^{\infty} \langle z, \phi_j \rangle \phi_{j+n}.$$

It is evident that $\| Q_n(z) \| = \| z \|$ for all n, so Q_n does not converge to 0 in the strong operator topology. On the other hand

$$| \langle Q_n(z), w \rangle | = | \langle z, P_n(w) \rangle | \leq \| z \| \, \| P_n(w) \|;$$

hence Q_n converges to 0 in the weak operator topology.   □

We are particularly interested in compact operators. For the theory of compact operators, the natural notion of convergence in $\mathcal{L}(\mathcal{H}, \mathcal{H}')$ will be convergence in norm. The collection of compact operators turns out to be a *closed* subspace of $\mathcal{L}(\mathcal{H}, \mathcal{H}')$ in this topology. We therefore prepare for this discussion by sketching the proof that $\mathcal{L}(\mathcal{H}, \mathcal{H}')$ is complete.

Theorem V.1.4. The complex vector space $\mathcal{L}(\mathcal{H}, \mathcal{H}')$ is a complete metric space with the distance function given by $d(L, M) = \| L - M \|$. Hence $\mathcal{L}(\mathcal{H}, \mathcal{H}')$ is a Banach space.

Proof. We provide a sketch and ask the reader to fill in the details. Suppose that $\{L_n\}$ is a Cauchy sequence in $\mathcal{L}(\mathcal{H}, \mathcal{H}')$. We define its limit

L by $L(z) = \lim L_n(z)$. The basic properties of limits guarantee that L is a well-defined linear transformation. Using the continuity of the norm on $\mathcal{H}'$ we obtain $\| L(z) \| \leq \lim \| L_n \| \, \| z \|$. Since the sequence of norms is Cauchy, we see that $\lim \| L_n \|$ exists and therefore L is bounded. $\square$

Exercise 1. Fill in the details of the proof of Theorem V.1.4.

Exercise 2. Let P_n and Q_n be as in Example V.1.3. Prove that $Q_n = P_n^*$.

V.2   Compact operators on Hilbert space

Compact operators on a Hilbert space are bounded linear transformations whose properties resemble those of finite-dimensional linear transformations. We will see that they are limits in norm of such transformations. Historically, compact operators first arose as integral operators used in solving differential equations. Because of the importance of differential equations in applied mathematics, the theory of compact operators is highly developed. We restrict our attention to compact operators on Hilbert spaces, although the concept of compact operator applies also to linear transformations between Banach spaces and has many uses in the Banach space setting.

Definition V.2.1. Suppose $L : \mathcal{H} \to \mathcal{H}'$ is a bounded linear transformation. Then L is called *compact* if, whenever $\{z_\nu\}$ is a bounded sequence in $\mathcal{H}$, the image sequence $\{L(z_\nu)\}$ has a convergent subsequence in $\mathcal{H}'$.

A compact linear transformation between Hilbert spaces is also called a *compact operator*. The word *compact* arose because of the analogy with the characterization of compactness in metric spaces in terms of convergent subsequences. An older term for compact operator is *completely continuous* operator.

It follows from the definition that a compact operator L is bounded, because the image of the closed unit ball under L must be a bounded set. The definition also implies that an operator with finite-dimensional range is compact. On the other hand, in infinite dimensions, the identity operator is not compact; this statement is evident because a sequence of orthonormal vectors has no convergent subsequence.

In Corollary V.2.4 we show that the collection of compact operators is a closed subspace of $\mathcal{L}(\mathcal{H}, \mathcal{H}')$. The special case where $\mathcal{H} = \mathcal{H}'$ is the most important; in this case we write $\mathcal{K}(\mathcal{H})$ for the space of compact operators.

Given our emphasis on inequalities, we wish to express the definition of compact operator as an inequality. We first make an elementary observation, commonly called the *small-constant large-constant trick*.

Lemma V.2.2. For each $\epsilon > 0$, there is a $C_\epsilon > 0$ such that, for all $x, y \in \mathbf{R}$,

$$|xy| \le \epsilon x^2 + C_\epsilon y^2.$$

Proof. The simple proof evokes the proof of the Cauchy-Schwarz inequality. For all x, y one has

$$0 \le \left(\sqrt{\epsilon}\, |x| - \frac{1}{2\sqrt{\epsilon}}\, |y| \right)^2 = \epsilon x^2 + \frac{1}{4\epsilon} y^2 - |xy|.$$

The result follows using $C_\epsilon = \frac{1}{4\epsilon}$. $\qquad\square$

A useful application of Lemma V.2.2 is the inequality

$$|\langle z, w \rangle| \le \epsilon \|z\|^2 + C \|w\|^2. \tag{1}$$

This version follows by combining the Cauchy-Schwarz inequality with Lemma V.2.2. The small-constant large-constant trick provides a useful formulation of compactness. We could restate the following result by absorbing the constant C into the definition of K; the statement here makes the crucial point more emphatically.

Proposition V.2.3. Let $L : \mathcal{H} \to \mathcal{H}'$ be linear. The following three statements are equivalent:

C1) L is compact.

C2) For each $\epsilon > 0$, there is a $C = C_\epsilon > 0$ and a compact operator $K = K_\epsilon$ such that

$$\| Lz \| \le \epsilon \| z \| + C \| Kz \|. \tag{2}$$

C3) For each $\epsilon > 0$, there is a $C = C_\epsilon > 0$ and a compact operator $K = K_\epsilon$ such that

$$\| Lz \|^2 \le \epsilon \| z \|^2 + C \| Kz \|^2. \tag{3}$$

Proof. First we will note that the two inequalities, with the same K but with different values of ϵ and C, are equivalent, and hence C2) and C3) are equivalent.

Suppose ϵ is given, and C3) holds. Write (3) with ϵ and C replaced by their squares to obtain

$$\| Lz \|^2 \le \epsilon^2 \| z \|^2 + C^2 \| Kz \|^2 \le (\epsilon \| z \| + C \| Kz \|)^2;$$

(2) follows by taking square roots.

Suppose $\epsilon > 0$ is given, and C2) holds. Choose η with $\epsilon = 2\eta^2$ and apply (2) with ϵ replaced by η. Squaring the result implies the estimate

$$\| Lz \|^2 \le \eta^2 \| z \|^2 + C^2 \| Kz \|^2 + 2\eta C \| z \| \, \| Kz \|.$$

Lemma V.2.2 implies the estimate

$$2\eta C \| z \| \, \| Kz \| \le \eta^2 \| z \|^2 + C' \| Kz \|^2.$$

Putting these together yields

$$\| Lz \|^2 \le 2\eta^2 \| z \|^2 + (C^2 + C') \| Kz \|^2 = \epsilon \| z \|^2 + C'' \| Kz \|^2$$

and hence (3) holds.

To prove the proposition it therefore suffices to prove that C2) is equivalent to compactness. When L is known to be compact, we choose $K = L$ and $C = 1$, and (2) holds for every positive ϵ.

The converse is more interesting. Let $\{z_\nu\}$ be a bounded sequence in $\mathcal{H}$. We want to extract a Cauchy subsequence from $\{L(z_\nu)\}$. From (2) we have

$$\| L(z_\nu) - L(z_\mu) \| = \| L(z_\nu - z_\mu) \|$$
$$\leq \epsilon \| z_\nu - z_\mu \| + C_\epsilon \| K_\epsilon (z_\nu - z_\mu) \|. \quad (4)$$

Given a positive integer n, we may choose ϵ sufficiently small in (4) so that the first term on the right-hand side is at most $\frac{1}{2n}$. The second term can then be made smaller than $\frac{1}{2n}$ by extracting a subsequence of the $\{z_\nu\}$ (still labeled the same) for which $\{K_\epsilon(z_\nu)\}$ converges, and then choosing ν and μ large enough.

Let $\{z^{(0)}\}$ denote the original bounded sequence. The above argument shows that, for each positive integer n, there is a sequence $\{z^{(n)}\}$, satisfying 1) and 2):

1) For $n \geq 1$, the sequence $\{z^{(n)}\}$ is a subsequence of $\{z^{(n-1)}\}$.

2) For any pair of terms ζ and w in $\{z^{(n)}\}$,

$$\| L(\zeta) - L(w) \| < \frac{1}{n}.$$

Let $\{w\}$ be the diagonal sequence defined by $w_k = z_k^{(k)}$. Then $\{w\}$ is a subsequence of $\{z^{(0)}\}$ and the image sequence under L of $\{w\}$ is Cauchy. Since $\mathcal{H}'$ is complete, the image sequence converges. Therefore L is a compact operator. $\qquad \square$

The proof of the next statement shows how to use Proposition V.2.3.

Corollary V.2.4. The compact operators form a *closed* subspace of $\mathcal{L}(\mathcal{H}, \mathcal{H}')$.

Proof. It is easy to see that the compact operators form a subspace. If L is compact, and c is a scalar, then cL is compact. The conclusion follows directly either from Definition V.2.1 or from Proposition V.2.3. Similarly, the sum of two compact operators is compact. Thus the compact operators form a subspace.

The main point is that this subspace is closed. Consider $T \in \mathcal{L}(\mathcal{H}, \mathcal{H}')$. Suppose, for each $\epsilon > 0$, there is a compact operator A_ϵ such that $\|T - A_\epsilon\| \leq \epsilon$. We will use Proposition V.2.3 to show that T is compact. This statement implies that the compact operators form a *closed* subspace of $\mathcal{L}(\mathcal{H}, \mathcal{H}')$.

For a given positive ϵ we may write, with $A = A_\epsilon$,

$$\|Tz\| = \|(T - A)z + Az\|$$

$$\leq \|(T - A)z\| + \|Az\| \leq \epsilon\|z\| + \|Az\|.$$

Proposition V.2.3 guarantees that T is compact. $\square$

The next result is a corollary of the proof of Proposition V.2.3. In Corollary V.2.5 and Proposition V.2.3, one can include the constants C_j with the compact operators K_j.

Corollary V.2.5. Suppose that, for each $\epsilon > 0$, there is a finite number of compact operators $K_1, \ldots, K_N$ and constants $C_1, \ldots, C_N$ such that

$$\|Lz\| \leq \epsilon\|z\| + \sum_{j=1}^{N} C_j \|K_j z\|.$$

Then L is compact.

Corollary V.2.6. Suppose that $L : \mathcal{H} \to \mathcal{H}'$ is compact, M_1 is bounded on $\mathcal{H}$, and M_2 is bounded on $\mathcal{H}'$. Then both LM_1 and M_2L are compact.

Proof. We write M in both cases, forgetting the subscript. To show that LM is compact we use the definition. Let $\{z_\nu\}$ be a bounded sequence in $\mathcal{H}$. Then $\{M(z_\nu)\}$ is also bounded, because M is a bounded operator. Since L is compact, $\{L(M(z_\nu))\}$ has a convergent subsequence. Thus LM is compact.

To show that ML is compact we use Proposition V.2.3. Given $\epsilon > 0$, set $\epsilon' = \epsilon/\| M \|$ and apply (2) (with ϵ replaced by ϵ') to L. The result is

$$\| MLz \| \leq \| M \| \, \| Lz \| \leq \| M \| (\epsilon' \| z \| + C' \| Lz \|)$$

$$= \epsilon \| z \| + C \| Lz \|,$$

and hence ML is compact. $\qquad\qquad\square$

We now consider the special case when the domain and target Hilbert spaces are the same. In this case $\mathcal{L}(\mathcal{H}, \mathcal{H})$ is an algebra. The operations are addition, scalar multiplication, and composition of operators. Our results so far enable us to conclude that $\mathcal{K}(\mathcal{H})$ is an *ideal* in this algebra.

Corollary V.2.7. The collection $\mathcal{K}(\mathcal{H})$ of compact operators forms a (two-sided) ideal in the algebra $\mathcal{L}(\mathcal{H}, \mathcal{H})$.

Proof. The collection of compact operators is closed under addition by Corollary V.2.4. It is closed under composition on both sides by bounded operators by Corollary V.2.6. Thus $\mathcal{K}(\mathcal{H})$ is an ideal in $\mathcal{L}(\mathcal{H}, \mathcal{H})$. $\qquad\qquad\square$

Theorem V.2.8. Let $L : \mathcal{H} \to \mathcal{H}'$ be compact. Then $L^* : \mathcal{H}' \to \mathcal{H}$ is compact.

Proof. Given $\epsilon > 0$, it follows from (1) following Lemma V.2.2 that there is a C such that

$$\| L^*z \|^2 = \langle L^*z, L^*z \rangle = \langle z, LL^*z \rangle \leq \epsilon \| z \|^2 + C \| LL^*z \|^2.$$

By Corollary V.2.7, LL^* is compact. Therefore L^* satisfies C3) from Proposition V.2.3, and hence it is compact. $\qquad\qquad\square$

Theorem V.2.9. Suppose $L : \mathcal{H} \to \mathcal{H}'$. Then L is compact if and only if L^*L is compact.

Proof. If L is compact, then L^*L is compact by Corollary V.2.6. Suppose L^*L is compact, and ϵ is given. Again by (1) following Lemma V.2.2, we have

$$\| Lz \|^2 = \langle Lz, Lz \rangle = \langle z, L^*Lz \rangle \le \epsilon \| z \|^2 + C \| L^*Lz \|^2.$$

Thus L satisfies C3) of Proposition V.2.3, and hence is compact.   □

Proposition V.2.10. Let $L : \mathcal{H} \to \mathcal{H}$ be an operator. Then L is compact if and only if there is a sequence $\{L_j\}$ of operators such that the range of each L_j is finite-dimensional and $\lim_j \| L_j - L \| = 0$.

Proof. An operator with finite-dimensional range must be compact, because any bounded sequence in a finite-dimensional Hilbert space has a convergent subsequence. By Corollary V.2.4 the limit in norm of compact operators is also compact.

The converse is left to the reader as Exercise 5.   □

We close this section by giving an example of a compact operator which we will use in Chapter VII.

Example V.2.11. Let Ω be a domain in complex Euclidean space, and let $\mathcal{H} = L^2(\Omega)$. Consider integral operators T defined by

$$(Tf)(z) = \int_\Omega K(z, w) f(w) \, dV(w).$$

The function K on $\Omega \times \Omega$ is called the *integral kernel* for T. In Theorem V.4.1 we give a property of K that guarantees $T : \mathcal{H} \to \mathcal{H}$ is a bounded operator. There are many circumstances when T is compact. For example, suppose that K is continuous and has compact support; it follows from Theorem V.4.2 that T is a compact operator.

Exercise 3. Suppose that $L : \mathcal{H} \to \mathcal{H}$ is positive definite and compact. Show that the (positive) square root of L is compact. Suggestion: Use (3) from Proposition V.2.3.

Exercise 4. Let B be a compact operator. Suppose that A and $B - A$ (and hence B) are nonnegative definite. We write $0 \leq A \leq B$ to express these hypotheses. Prove that A is also compact. Suggestion: Use Proposition V.2.3 to first show that $\sqrt{A}$ is compact.

Exercise 5. Complete the proof of Proposition V.2.10. Recall that $\mathcal{H}$ is assumed to be separable. Suggestion: Let $\{v_j\}$ be a complete orthonormal system, and define L_n by

$$L_n(z) = \sum_{j=1}^{n} \langle z, v_j \rangle L(v_j).$$

Show that $\| L_n - L \|$ tends to 0. This idea also appears in the proof of Theorem V.3.5.

Exercise 6. Give an example of a sequence of compact operators converging in the strong operator topology to a limit that is not compact.

V.3 The spectral theorem for compact Hermitian operators

Before we proceed to the spectral theorem, we give a simple example of an operator on a Hilbert space that has no eigenvectors at all. This example indicates that spectral theory in infinite-dimensional Hilbert spaces must be more intricate than it is in finite dimensions. See [Ru] for general results. Here we will be content to study compact Hermitian operators, which behave remarkably like finite-dimensional operators.

Example V.3.1. Let $\mathcal{H} = l^2$, and define $L : \mathcal{H} \to \mathcal{H}$ by $L(a_1, a_2, \ldots) = (0, a_1, a_2, \ldots)$. Then $Lz = \lambda z$ implies that $z = 0$, so

L has no eigenvectors. (L is Q_1 from Example V.1.3.) A simple modification gives a compact operator T with no eigenvectors. Put $T(a_1, a_2, \ldots) = (0, a_1, \frac{a_2}{2}, \ldots, \frac{a_j}{j}, \ldots)$.

In contrast with Example V.3.1, compact Hermitian operators have many eigenvectors. Some general remarks about eigenspaces help prepare us for the proof. Let $L : \mathcal{H} \to \mathcal{H}$ be a linear transformation and suppose λ is an eigenvalue for L. The *eigenspace E_λ* corresponding to λ is the subspace of $\mathcal{H}$ consisting of all z for which $Lz = \lambda z$. It is evident that an eigenspace is a linear subspace of $\mathcal{H}$. For Hermitian operators the eigenspaces corresponding to distinct eigenvectors are orthogonal (Lemma V.3.2), and for compact operators the eigenspaces must be finite-dimensional (Lemma V.3.3).

Lemma V.3.2. Suppose $L : \mathcal{H} \to \mathcal{H}$ and $L = L^*$. The eigenspaces of L corresponding to distinct eigenvalues are orthogonal.

Proof. Suppose that $Lz = \lambda z$ and $Lw = \mu w$. Recall that λ and μ are real. We have

$$\lambda \langle z, w \rangle = \langle Lz, w \rangle = \langle z, Lw \rangle = \mu \langle z, w \rangle.$$

Therefore $0 = (\lambda - \mu) \langle z, w \rangle$ and the result follows.   □

Lemma V.3.3. Suppose $L : \mathcal{H} \to \mathcal{H}$ and L is compact. Let E_λ be an eigenspace of L. Then either $\lambda = 0$ or E_λ is finite-dimensional.

Proof. Suppose that E_λ is infinite-dimensional. We can then choose a countably infinite orthonormal collection $\{z_j\}$ of elements in E_λ. Since these are unit vectors, they form a bounded sequence in $\mathcal{H}$. Since L is compact, the sequence $\{L(z_j)\}$ has a convergent subsequence; hence $\{\lambda z_j\}$ has a convergent subsequence. When $\lambda = 0$ this is perfectly fine; when $\lambda \neq 0$ however, this gives a contradiction. After dividing by λ we obtain a convergent subsequence of orthogonal unit vectors. Such a sequence cannot converge, because the distance between orthogonal unit vectors is $\sqrt{2}$. Therefore $\lambda = 0$ when E_λ is infinite-dimensional.

$\square$

Lemma V.3.4. Let $L : \mathcal{H} \to \mathcal{H}$ be compact and Hermitian. Suppose there is a linearly independent sequence of eigenvectors for L with corresponding eigenvalues $\lambda_1, \lambda_2, \ldots$. Suppose further that $\lambda = \lim_{j\to\infty} \lambda_j$ exists. Then $\lambda = 0$.

Proof. For each j there is a unit vector z_j for which $L(z_j) = \lambda_j z_j$. These z_j form a bounded sequence in $\mathcal{H}$; since L is compact, the sequence $\{L(z_j)\}$ has a convergent subsequence, denoted by $\{L(w_k)\}$. The limit ζ of $\{L(w_k)\}$ exists, and the limit λ of $\{\lambda_k\}$ exists. If $\lambda \neq 0$, then $\{w_k\}$ converges to $\frac{\zeta}{\lambda}$. By Lemma V.3.3, each eigenspace is finite-dimensional, so $\{w_k\}$ is not eventually constant. On the other hand, by Lemma V.3.2, the eigenspaces are orthogonal. As in the last step of the proof of Lemma V.3.3, a sequence of orthogonal unit vectors cannot converge. We obtain a contradiction unless $\lambda = 0$. $\square$

Let $L:\mathcal{H} \to \mathcal{H}$ be a Hermitian operator. As in Chapter IV, we consider the Hermitian form $z \to \langle Lz, z \rangle$, and apply constrained optimization. The following two measurements of the size of L turn out to be equal:

$$\| L \| = \sup_{z\neq 0} \frac{\| Lz \|}{\| z \|}$$

$$\| \| L \| \| = \sup_{z\neq 0} \frac{|\langle Lz, z \rangle|}{\| z \|^2}.$$

For use in the next proof we note, by Proposition IV.1.5, that $\| \| L \| \| \neq 0$ when $L \neq 0$.

When L is both compact and Hermitian we can say much more. Our first result finds one eigenvalue, and prepares us for the spectral theorem.

Theorem V.3.5. (Optimization for a compact Hermitian operator) Let $L:\mathcal{H} \to \mathcal{H}$ be a Hermitian operator. Then $\| L \| = \| \| L \| \|$.

Suppose in addition that L is compact. Then there is an eigenvalue λ with $|\lambda| = \|L\|$.

Proof. When L is the zero operator the conclusion holds; we assume that $L \neq 0$. Thus we may divide by $\|L\|$ or $\||L|\|$ if necessary.

To verify the equality $\|L\| = \||L|\|$, we prove two inequalities. One is easy; $|\langle Lz, z \rangle| \leq \|Lz\| \|z\|$ by the Cauchy-Schwarz inequality. Therefore

$$\||L|\| = \sup_{z \neq 0} \frac{|\langle Lz, z \rangle|}{\|z\|^2} \leq \sup_{z \neq 0} \frac{\|Lz\|}{\|z\|} = \|L\|. \tag{5}$$

The converse inequality uses a nice polarization trick. We begin with the identity

$$4 \operatorname{Re}\langle Lz, w \rangle = 2\langle Lz, w \rangle + 2\langle Lw, z \rangle$$
$$= \langle L(z + w), z + w \rangle - \langle L(z - w), z - w \rangle.$$

From this identity we obtain

$$4 \operatorname{Re}\langle Lz, w \rangle \leq \||L|\| \left(\|z + w\|^2 + \|z - w\|^2 \right). \tag{6}$$

Applying the parallelogram law to (6) then gives

$$4 \operatorname{Re}\langle Lz, w \rangle \leq \||L|\| \left(2\|z\|^2 + 2\|w\|^2 \right). \tag{7}$$

Inequality (7) holds for all z and w. We choose w to equal $\frac{Lz}{\||L|\|}$ in (7) and obtain

$$4 \frac{\|Lz\|^2}{\||L|\|} \leq \||L|\| \left(2\|z\|^2 + 2\frac{\|Lz\|^2}{\||L|\|^2} \right). \tag{8}$$

Simplifying (8) yields

$$\|Lz\|^2 \leq \||L|\|^2 \|z\|^2.$$

After dividing both sides by $\|z\|^2$ and taking the supremum over nonzero z, we obtain $\|L\|^2 \leq \||L|\|^2$. Combining this inequality with the opposite inequality (5) completes the proof of the first part.

Next we invoke compactness. Choose a sequence $\{z_\nu\}$ on the unit sphere such that $|\langle Lz_\nu, z_\nu \rangle|$ converges to $|||\,L\,|||$ and hence to $||\,L\,||$. We can find a subsequence, still labeled $\{z_\nu\}$, such that $\langle Lz_\nu, z_\nu \rangle$ converges to either $\pm||\,L\,||$. We write μ instead of $\pm||\,L\,||$. Since $\{z_\nu\}$ is a bounded sequence, and L is compact, it follows that there is a subsequence, again labeled $\{z_\nu\}$, for which $\{L(z_\nu)\}$ converges to some w. We claim that $||\,w\,|| = ||\,L\,||$ and also that w is in the range of L.

First we show that $||\,w\,|| = ||\,L\,||$. Since $|\langle Lz_\nu, z_\nu \rangle| \to ||\,L\,||$, we also have $|\langle w, z_\nu \rangle| \to ||\,L\,||$. Since z_ν is a unit vector the Cauchy-Schwarz inequality forces $||\,w\,|| \geq ||\,L\,||$; on the other hand $||\,Lz_\nu\,|| \leq ||\,L\,||$ by definition of $||\,L\,||$. Therefore $||\,w\,|| \leq ||\,L\,||$ as well. Combining the two inequalities gives $||\,w\,|| = ||\,L\,||$.

Recall that $\{L(z_\nu)\}$ converges to w. We need to show that $\{z_\nu\}$ converges; we will show that μz_ν converges to w. To do so, it suffices to show that $||\,L(z_\nu) - \mu z_\nu\,||$ converges to zero. Expanding the squared norm gives

$$||\,L(z_\nu) - \mu z_\nu\,||^2 = \left|\left|\,L(z_\nu)\,\right|\right|^2 - 2\,\mathrm{Re}(\mu\langle Lz_\nu, z_\nu\rangle) + ||\,L\,||^2. \quad (9)$$

Since $\langle Lz_\nu, z_\nu \rangle \to \mu$, and $||\,w\,|| = ||\,L\,||$, the right-hand side of (9) tends to

$$||\,w\,||^2 - 2\mu^2 + ||\,L\,||^2 = 0.$$

Hence μz_ν converges to w. Put $z = \frac{w}{\mu}$. Then z_ν converges to the unit vector z and $L(z_\nu)$ converges to w. Therefore

$$Lz = \lim L(z_\nu) = w = \mu z = \pm||\,L\,||z,$$

and hence z is the desired eigenvector. $\qquad\qquad\square$

The equality $||\,L\,|| = |||\,L\,|||$ for a Hermitian operator has a nice consequence.

Corollary V.3.6. Let $T : \mathcal{H} \to \mathcal{H}$ be a bounded operator. Then $||\,T\,||^2 = ||\,T^*T\,||$.

Proof. Since T^*T is Hermitian, we have

$$\| T^*T \| = \| \| T^*T \| \| = \sup_{\| z \| = 1} \langle T^*Tz, z \rangle$$
$$= \sup_{\| z \| = 1} \| Tz \|^2 = \| T \|^2. \tag{10}$$

$\square$

We may now mimic the proof of the spectral theorem in finite dimensions by finding new (unit) eigenvectors orthogonal to the previously found eigenvectors. This procedure will produce a (possibly finite) sequence of eigenvalues, whose absolute values are nonincreasing.

Theorem V.3.7. (Spectral Theorem for compact Hermitian operators) Suppose that $L : \mathcal{H} \to \mathcal{H}$ is compact and Hermitian. Then there is a complete orthonormal system in $\mathcal{H}$ consisting of eigenvectors of L.

Proof. In Theorem V.3.5 we found one eigenvector z_1 with corresponding eigenvalue λ_1 whose absolute value is maximal. Also we have $\| z_1 \| = 1$. Let $\mathcal{H}_1$ be the orthogonal complement of the span of z_1. Notice that $\langle w, z_1 \rangle = 0$ implies that

$$\langle Lw, z_1 \rangle = \langle w, Lz_1 \rangle = \lambda_1 \langle w, z_1 \rangle = 0.$$

Therefore the restriction of L to $\mathcal{H}_1$ maps $\mathcal{H}_1$ to itself. This restriction is also compact and Hermitian. We apply Theorem V.3.5 to this restriction, obtaining an eigenvector z_2 with eigenvalue λ_2 of maximal absolute value. Also z_2 is orthogonal to z_1; we may assume that it is a unit vector.

Proceeding in this way we obtain an orthonormal set of eigenvectors $\{z_j\}$ for L with corresponding eigenvalues λ_j. We also have $|\lambda_{j+1}| \leq |\lambda_j|$.

Each eigenspace E_λ for $\lambda \neq 0$ is finite-dimensional by Lemma V.3.3. By Lemma V.3.4 the only possible cluster point of the sequence of eigenvalues is 0. Therefore, either $\mathcal{H}$ is itself finite-dimensional, or $\lim(\lambda_j) = 0$. In any case, the sequence of numbers $|\lambda_j|$ has 0 as its only possible limit point.

Consider a maximal collection M of orthonormal eigenvectors. Let P_n denote the projection operator defined by

$$P_n(\zeta) = \sum_{j=1}^{n} \langle \zeta, z_j \rangle z_j.$$

Notice, for $\zeta \in \mathcal{H}$, that

$$\| L(P_n(\zeta)) - L(\zeta) \| \leq |\lambda_{n+1}| \, \| \zeta \|.$$

The absolute values of the eigenvalues are decreasing to zero with n; it follows that $L(P_n \zeta)$ converges to $L\zeta$ as n tends to infinity. For $w \in R(L)$ (the range of L), we therefore have

$$w = L(\zeta) = \lim L(P_n(\zeta)) = \lim_{n \to \infty} \sum_{j=1}^{n} \langle \zeta, z_j \rangle \lambda_j z_j.$$

Therefore each vector in $R(L)$ can be expanded in terms of elements of M. Next note that $N(L)$ is the 0-eigenspace of L. Thus each $v \in N(L)$ has an expansion in terms of those eigenvectors in M that are also in $N(L)$.

Since $L = L^*$, it follows from Proposition IV.2.2 that $N(L) \oplus R(L)$ is dense in $\mathcal{H}$. The vectors in M span both $N(L)$ and $R(L)$; since the sum of these subspaces is dense, the zero vector is the only vector orthogonal to M. Therefore the orthonormal system M of eigenvectors is complete. □

Application V.3.8. We use Theorem V.3.5 to prove the following remarkable inequality. Assume that $g(0) = 0$, and that g is continuously differentiable on $[0, 1]$. Write $\| g \|^2$ for the usual squared L^2 norm

$$\| g \|^2 = \int_0^1 |g(t)|^2 \, dt.$$

Then we have

$$\| g \| \le \frac{2}{\pi} \| g' \|,$$

and the constant $\frac{2}{\pi}$ is best possible.

Proof. Define the operator T (the Volterra integral operator) by

$$Tf(x) = \int_0^x f(t)\, dt.$$

It is easy to show (Exercise 7) that T is compact. By interchanging the order of integration in the double integral defining $\langle Tf, h \rangle$, we see that

$$T^* f(y) = \int_y^1 f(t)\, dt$$

and hence that

$$TT^* f(x) = \int_0^x \int_y^1 f(t)\, dt\, dy.$$

Since T is compact, so is T^*T by Corollary V.2.6. The operator TT^* is therefore both compact and Hermitian. By Theorem V.3.5, there is an eigenvector λ for which $|\lambda| = \| TT^* \|$. We can find all eigenvalues of TT^* by solving the integral equation

$$\int_0^x \int_y^1 f(t)\, dt\, dy = \lambda f(x).$$

To do so we differentiate twice using the fundamental theorem of calculus to obtain the differential equation

$$-f(x) = \lambda f''(x)$$

and the boundary conditions $f'(1) = f(0) = 0$. We solve the equation by elementary methods to get the eigenvalues and eigenvectors. For each nonnegative integer n there is an eigenvalue λ with

$$\frac{1}{\sqrt{\lambda}} = \frac{\pi}{2} + n\pi$$

and a corresponding eigenfunction f given by

$$f(x) = \sin\left(\frac{(2n+1)\pi x}{2}\right).$$

The maximum eigenvalue occurs when $n = 0$, and we see that $\sqrt{\lambda} = \frac{2}{\pi}$. Thus $|| T ||^2 = || TT^* || = \frac{4}{\pi^2}$ and $|| T || = \frac{2}{\pi}$. For all $f \in L^2([0, 1])$, we therefore have $|| Tf || \leq \frac{2}{\pi} || f ||$. When g is continuously differentiable and $g(0) = 0$, we have $g = T(g')$ and hence

$$|| g || = || T(g') || \leq \frac{2}{\pi} || g' ||.$$

Since equality occurs when $g(x) = \sin(\frac{\pi x}{2})$, the constant $\frac{2}{\pi}$ is best possible. □

Remark V.3.9. (Versions of Wirtinger's inequality) The inequality from Application V.3.8 is another version of Wirtinger's inequality. Suppose f is continuously differentiable, periodic of period 2π, and its average value on $[0, 2\pi]$ is zero. In Application IV.7.16 we showed using Fourier series that

$$\int_0^{2\pi} | f(x) |^2 \, dx \leq \int_0^{2\pi} | f'(x) |^2 \, dx. \tag{11}$$

Equality holds only for linear combinations of cosine and sine. In Application V.3.8 we assume that $f(0) = 0$. In Application IV.7.16 we assumed that the average value of f is zero, and in Application IV.7.17 we gave a related inequality under the assumption that $f \in L^1$. Some assumption on the values of f is required; by adding a constant to f we can increase the left-hand side of (11) without changing the right-hand side. Assuming that the average value of f is zero amounts to assuming that f is orthogonal to the constant functions.

Exercise 7. Prove that T as defined in Application V.3.8 is compact.

Exercise 8. Use Application V.3.8 to derive Application IV.7.16.

V.4   Integral operators

In this section we suppose that $\mathcal{H} = L^2(\Omega)$, where Ω is a nice sub-set (such as open or closed) of real (or complex) Euclidean space. We often define linear transformations on $\mathcal{H}$ by integration with respect to an integral kernel. Such transformations are known as *integral operators*; they have been systematically studied because they arise in the solutions of linear differential equations. Important properties (such as compactness) for general operators were often first introduced for integral operators.

Let Ω be a nice subset of real or complex Euclidean space. Let $K : \Omega \times \Omega \to \mathbf{C}$ be a continuous function. We define an operator T by

$$(Tf)(x) = \int_\Omega K(x, y) f(y) \, dV(y). \tag{12}$$

For each scalar c we have $T(cf) = cT(f)$, and for each f and g we have $T(f + g) = Tf + Tg$. Thus T is a linear transformation from some domain to some target. In general it is difficult to determine the precise spaces between which T is a bounded operator. One favorable case is when there is a constant C such that, for all x and y in Ω, both (13) and (14) hold:

$$\int_\Omega | K(x, y) | \, dV(x) \le C \tag{13}$$

$$\int_\Omega | K(x, y) | \, dV(y) \le C. \tag{14}$$

In this case T maps $L^2(\Omega)$ to $L^2(\Omega)$; furthermore T is a bounded linear transformation whose operator norm is at most C. This statement is our next result. The version here is adequate for our purposes, but can be easily generalized. For example, the *integral kernel K* need not be continuous, and we may replace L^2 with other spaces. See [F].

Theorem V.4.1. Suppose that K is continuous and there is a constant C such that (13) and (14) hold. Formula (12) then defines an integral operator T; it is a bounded linear transformation from $L^2(\Omega)$ to $L^2(\Omega)$, and $\| T \| \leq C$.

Proof. First we estimate $| Tf(x) |$ using the Cauchy-Schwarz inequality:

$$| Tf(x) | \leq \int_\Omega | K(x, y) || f(y) | \, dV(y)$$

$$= \int_\Omega | K(x, y) |^{\frac{1}{2}} | K(x, y) |^{\frac{1}{2}} | f(y) | \, dV(y)$$

$$\leq \left(\int_\Omega | K(x, y) | \, dV(y) \right)^{\frac{1}{2}} \left(\int_\Omega | K(x, y) | | f(y |^2 \, dV(y) \right)^{\frac{1}{2}}.$$

Therefore

$$| Tf(x) |^2 \leq \int_\Omega | K(x, y) | \, dV(y) \int_\Omega | K(x, y) | | f(y) |^2 \, dV(y)$$

$$\leq C \int_\Omega | K(x, y) | | f(y) |^2 \, dV(y). \tag{15}$$

Integrate (15) with respect to x. We may interchange the order of integration on the right-hand side, because the integrand is nonnegative. We then obtain

$$\| Tf \|^2 = \int_\Omega | Tf(x) |^2 \, dV(x) \leq C^2 \int_\Omega | f(y) |^2 \, dV(y) = C^2 \| f \|^2,$$

which finishes the proof. $\qquad\qquad\qquad\Box$

In the rest of this chapter we write $| x |$ for the Euclidean norm of x to avoid confusion with the L^2 norm of a function. Let W be a closed and bounded (hence compact) subset of real (or complex) Euclidean space, and let K be a continuous complex-valued function on $W \times W$. Because $W \times W$ is a compact set, a standard theorem in analysis guarantees that K is uniformly continuous. Hence, for each $\epsilon > 0$, there is a $\delta > 0$ such that $|x - x'| < \delta$ and $|y - y'| < \delta$ together imply

$$| K(x, y) - K(x', y') | < \epsilon.$$

Also $|K|$ is bounded on $W \times W$.

In this case Theorem V.4.1 applies, and T (defined in (12) with Ω replaced by W) will be a bounded operator on $L^2(W)$. In fact much more is true. We gather this information in the next result.

Theorem V.4.2. Suppose $W \subset \mathbf{R}^n$ is compact. Let K be continuous on $W \times W$, and let T be the operator with integral kernel K. The following statements are true:

1) T is a bounded operator on $L^2(W)$.

2) If $f \in L^2(W)$, then Tf is continuous on W.

3) Suppose that $\{f_n\}$ is a bounded sequence in $L^2(W)$. Then the collection of functions $\{Tf_n\}$ is equicontinuous and uniformly bounded.

4) T is a compact operator on $L^2(W)$.

Proof.

1) Since W is compact, $W \times W$ is also compact. Since K is continuous on the compact set $W \times W$, it is bounded; suppose $|K(x, y)| \leq m$ for all x and y. Both (13) and (14) hold with $C = m\mathrm{Vol}(W)$. Thus Theorem V.4.1 applies, and 1) is true.

2) Suppose that $x, x' \in W$. We compute

$$| Tf(x) - Tf(x') | = \left| \int_W (K(x, y) - K(x', y)) f(y)\, dV(y) \right|$$

$$\leq \int_W | K(x, y) - K(x', y) | \, | f(y) |\, dV(y).$$

$$(16)$$

We estimate the right-hand side of (16) using the Cauchy-Schwarz inequality and then square to obtain

$$| Tf(x) - Tf(x') |^2 \leq \int_W | K(x, y) - K(x', y) |^2 dV(y) \, \| f \|^2$$
$$\leq \text{Vol}(W) \sup_{x,x',y} | K(x, y)$$
$$- K(x', y) |^2 \| f \|^2. \tag{17}$$

Suppose $\epsilon > 0$ is given. Since K is uniformly continuous, there is a $\delta > 0$ such that $| x - x' | < \delta$ implies $| K(x, y) - K(x', y) |^2$ is as small as we wish, independently of y. So we may choose δ to ensure that the right-hand side of (17) is at most ϵ^2. Therefore Tf is continuous at x'. Since x' is arbitrary, Tf is continuous.

3) The proof of equicontinuity is virtually the same. Let $\{f_n\}$ be a bounded sequence. By (17) we have

$$| Tf_n(x) - Tf_n(x') |$$
$$\leq \left(\int_W | K(x, y) - K(x', y) |^2 dV(y) \right)^{\frac{1}{2}} \| f_n \|. \tag{18}$$

Since $\{\| f_n \|\}$ is a bounded set of numbers and W is compact, we may choose δ independent of n to ensure that (18) is less then ϵ for all n whenever $| x - x' | < \delta$. This proves equicontinuity.

The proof that the functions Tf_n are uniformly bounded is similar and easier, so we leave it as Exercise 9.

4) To show that T is compact, we let $\{f_n\}$ be a bounded sequence in $L^2(W)$. By 3), the collection $\mathcal{F}$ of continuous functions $\{Tf_n\}$ is equicontinuous and uniformly bounded. By the Arzelà-Ascoli theorem (see the Appendix, [F], or [Kr]), $\mathcal{F}$ has compact closure in the metric space $C(W)$. Hence there is a subsequence $\{Tf_{n_k}\}$ that converges uniformly. Since W is bounded, uniform convergence guarantees that $\{Tf_{n_k}\}$ converges in $L^2(W)$. We have proved that T is compact on $L^2(W)$.   □

Remark V.4.3. We note also a simple way to show that certain operators are Hermitian. Let T be an integral operator on $A^2(\Omega)$, where $\Omega \subset \mathbf{C}^n$, and suppose that T has integral kernel $K(z, \overline{w})$. Then T is Hermitian if K satisfies $K(z, \overline{w}) = \overline{K(w, \overline{z})}$. See Lemma III.3.2 for the case where T is the Bergman projection and see Proposition VI.2.10 for another instance.

Remark V.4.4. Suppose that the integral kernel K of an operator T lies in $L^2(\Omega \times \Omega)$. Then T is compact. One way to prove this conclusion is to express T as a limit of finite-dimensional operators and apply Proposition V.2.10.

Remark V.4.5. Another approach to Theorem V.4.2 uses uniform approximation of continuous functions by polynomials (Exercise 28 from Chapter IV) and Proposition V.2.10.

Exercise 9. Show that the functions Tf_n from Theorem V.4.2 are uniformly bounded.

Exercise 10. In each case show that the operator T is not compact on $L^2([0, 1])$:

$$Tf(x) = \frac{1}{x} \int_0^x f(t) \, dt$$
$$Tf(x) = xf(x).$$

Suggestion for the second: Look for eigenfunctions, and contradict the conclusion of the spectral theorem.

V.5   A glimpse at singular integral operators

In this section we give some examples of integral operators whose analysis requires deeper ideas than we cover in this book. The material is not required in the rest of the book, although some of the examples help explain the proof of Theorem VII.1.1.

First we give an example of an unbounded operator with an interval of eigenvalues. Then we discuss the Hilbert transform and show that certain related commutators define compact operators. We discuss the Fourier transform and sketch the definition of pseudodifferential operator. This material provides a nice approach to the appealing notion of fractional derivatives and integrals.

Example V.5.1. **(An interval of eigenvalues)** Let V denote the space of bounded continuous complex-valued functions on **R**. Consider the linear transformation T defined on V by

$$Tf(x) = \int_{\mathbf{R}} e^{-|x-t|} f(t) \, dt.$$

The integral kernel $e^{-|x-t|}$ is continuous and symmetric in x and t. In case $f \in L^2(\mathbf{R})$, the Cauchy-Schwarz inequality shows that Tf is defined; in general however Tf will not be in $L^2(\mathbf{R})$. Suppose the domain of T is the subspace of $f \in L^2$ for which $Tf \in L^2$. Then T becomes a good example of an unbounded operator.

Let us consider T on V. For each $a \geq 0$, we show that the function f_a given by $f_a(x) = e^{-iax}$ is an eigenfunction for T with corresponding eigenvalue $\frac{2}{1+a^2}$:

$$\int_{-\infty}^{\infty} e^{-|x-t|-iat} \, dt = \int_{-\infty}^{x} e^{-x+t-iat} \, dt + \int_{x}^{\infty} e^{x-t-iat} \, dt$$

$$= e^{-x}\frac{e^{x(1-ia)}}{1-ia} + e^{x}\frac{e^{-x(1+ia)}}{1+ia}$$

$$= e^{-iax}\left(\frac{1}{1-ia} + \frac{1}{1+ia}\right) = e^{-iax}\frac{2}{1+a^2}.$$

Thus we have

$$T(f_a) = \frac{2}{1+a^2} f_a.$$

The functions e^{-iax} are not square integrable. If T is defined on V, then each point in the interval $(0, 2]$ is an eigenvalue for T. The reader should contrast this situation with the point spectrum of a compact Hermitian operator described in Section V.4.

Next we turn to the Hilbert transform. We barely scratch the surface of what is known. Recall that the collection of exponentials $\{e^{int}\}$ for $n \in \mathbf{Z}$ forms a complete orthonormal system for $L^2(S^1)$. We define a projection $P : L^2(S^1) \rightarrow L^2(S^1)$ by orthogonally projecting onto the span of the e^{int} for $n \geq 0$. Thus $P(e^{imt}) = 0$ for $m < 0$, and $P(e^{imt}) = e^{imt}$ otherwise. Then $P = P^* = P^2$.

Definition V.5.2. The Hilbert transform $H : L^2(S^1) \rightarrow L^2(S^1)$ is defined by $H = 2P - I$.

Lemma V.5.3. The Hilbert transform is Hermitian and unitary.

Proof. Since $P^* = P$ it follows that $H^* = H$ and also that

$$H^2 = (2P - I)^2 = 4P^2 - 4P + I = I;$$

hence $H^{-1} = H$ and $H = H^*$. Therefore H is unitary. $\qquad\square$

After we recall the notion of principal value integral, we will give an equivalent definition of H as a singular integral operator.

Let f be continuous on an interval $[a, b]$ except for a single point c in the interior. Suppose we wish to compute $\int_a^b f(x)\,dx$, but this integral does not exist. A simple example is provided by $\int_{-1}^1 \frac{1}{x}\,dx$. By symmetry the limit

$$\lim_{\epsilon \to 0} \int_{\epsilon \leq |x| \leq 1} \frac{1}{x}\,dx$$

does exist. We therefore consider the following symmetric limit, which is called the *Cauchy Principal Value*:

$$PV \int_a^b f(x)\,dx = \lim_{\epsilon \to 0} \left(\int_a^{c-\epsilon} + \int_{c+\epsilon}^b \right) f(x)\,dx.$$

To discuss the Hilbert transform we need to extend the definition of Cauchy principal value to line integrals. For simplicity we work on the unit circle S^1.

Definition V.5.4. Suppose that u is continuous on S^1 and fix $e^{it} \in S^1$. We define the Cauchy principal value integral

$$PV \int_{S^1} \frac{u(\zeta)}{\zeta - e^{it}}\,d\zeta$$

by

$$PV \int_{S^1} \frac{u(\zeta)}{\zeta - e^{it}}\,d\zeta = \lim_{\epsilon \to 0} \int_{S^1 - B_\epsilon(e^{it})} \frac{u(\zeta)}{\zeta - e^{it}}\,d\zeta,$$

where $B_\epsilon(e^{it})$ denotes a symmetric interval of length 2ϵ about e^{it}.

We state the following formula of Hilbert; although the proof follows easily from the Cauchy integral formula, we omit it here.

Theorem V.5.5. Let u be continuous on S^1. Then

$$(Hu)(e^{it}) = \frac{1}{\pi i} PV \int_{S^1} \frac{u(\zeta)}{\zeta - e^{it}}\,d\zeta.$$

Recall that the *commutator* $[A, B]$ of operators A and B is the operator $AB - BA$. The following lemma and proposition arise in the study of singular integrals, and are related to Theorem VII.1.1. See also Lemma VII.6.5 and Remark VII.6.6.

Lemma V.5.6. Let M denote multiplication by a trigonometric polynomial on $L^2(S^1)$. Let $P = \frac{H+I}{2}$ be the projection defined before Definition V.5.2. Then the commutator $[M, P]$ has finite-dimensional range and hence is a compact operator on $L^2(S^1)$.

Proof. It suffices to prove the result when M is the operator M_k given by multiplication by e^{ikt}. Suppose f in $L^2(S^1)$ has the orthonormal expansion $\sum a_n e^{int}$. Then

$$[M_k, P](f) = M_k(Pf) - P(M_k f)$$

$$= \sum_{n \geq 0} a_n e^{i(n+k)t} - \sum_{n \geq -k} a_n e^{i(n+k)t} = \sum_{n=-k}^{-1} a_n e^{i(n+k)t}.$$

(When $k < 0$ the sum on the right-hand side is in an unusual order.) In any case this sum is a finite linear combination of linearly independent exponentials, so the range of the commutator is finite-dimensional. Therefore the commutator is compact. $\qquad\square$

Proposition V.5.7. Suppose that q is continuous on S^1. Let M_q denote multiplication by q on $L^2(S^1)$. Then the commutator $[M_q, P]$ is a compact operator on $L^2(S^1)$.

Proof. By Fejér's Theorem (Theorem IV.7.7) there is a sequence $\{q_n\}$ of trigonometric polynomials converging uniformly to q on S^1. By Lemma V.5.6, each $[M_{q_n}, P]$ is a compact operator. We want to show that $[M_q, P]$ is compact. By Corollary V.2.4 it suffices to show that $\| [M_q - M_{q_n}, P] \|$ converges to 0. Writing $h_n = q - q_n$, we see that h_n converges to zero uniformly, and we want to show $\| [M_{h_n}, P] \|$ converges to zero. This is easy; each term $\| M_{h_n} P \|$ and $\| P M_{h_n} \|$ is easily seen to converge to zero, because h_n converges uniformly to zero and P is a bounded operator. $\qquad\square$

Corollary V.5.8. With q as in Proposition V.5.7, the commutator $[H, M_q]$ is a compact operator.

Proof. By Proposition V.5.7 and the relationship between P and H we see that $[\frac{H+I}{2}, M_q]$ is compact. Since $[I, M_q] = 0$, we obtain the desired conclusion. $\qquad\square$

We now turn to the Fourier transform and give a brief introduction to the notion of pseudodifferential operator.

Example V.5.9. (**Fractional integrals and derivatives**) First assume that $a > 0$. We define an integral operator T_a on the space $C(\mathbf{R})$ by

$$(T_a f)(x) = \frac{1}{\Gamma(a)} \int_0^x (x - y)^{a-1} f(y) \, dy.$$

In case $a = 1$ we see that Tf is the integral of f. By induction, we see when a is a nonnegative integer that T_a integrates f precisely a times. Suppose that $a, b > 0$. Exercise 11 uses the Euler beta function to verify that the composition formula $T_b T_a = T_{b+a}$ is valid. Because of this property we are entitled to think of T_a as integrating a times, for an arbitrary positive real number a. Hence we can define *fractional integrals*.

Suppose that f is continuously differentiable. It is easy to verify by integration by parts that $T_a(f) = T_{a+1}(f')$. Now suppose that $-1 < a \le 0$. We define T_a on the space of continuously differentiable functions by

$$T_a(f) = T_{a+1}(f').$$

By induction we see how to define, for all real a, the operator T_a on the space of infinitely differentiable functions. We thus obtain *fractional derivatives*.

Mathematicians have developed general methods to analyze ideas such as fractional derivatives. These methods are beyond the scope of this book, but we will say a few words about one of them nonetheless. We sketch the definition of a class of pseudodifferential operators, using the Fourier transform as the basic tool. See [SR], [S], or [Ta] for more information.

Let $\mathcal{F}$ denote the Fourier transform, defined on $L^1(\mathbf{R}^n)$ by

$$\mathcal{F}(f)(\xi) = f^\wedge(\xi) = \left(\frac{1}{2\pi} \right)^{\frac{n}{2}} \int_{\mathbf{R}^n} e^{-ix \cdot \xi} f(x) \, dx.$$

For $f \in L^2(\mathbf{R}^n)$, this integral need not converge absolutely. Nevertheless it is possible to define the Fourier transform on $L^2(\mathbf{R}^n)$. The ideas are similar to those in Section IV.7. As in the case of Fourier series, the definition of $\mathcal{F}$ extends to various dual spaces via the formula $\mathcal{F}(g)(h) = g(\mathcal{F}(h))$. Using this approach, one can define $\mathcal{F}$ on spaces that contain $L^2(\mathbf{R}^n)$. The Plancherel theorem states that (the extended operator) $\mathcal{F}$ defines a unitary operator on $L^2(\mathbf{R}^n)$. An important step in the proof of the Plancherel Theorem is the Fourier inversion formula, stated in Theorem V.5.10 below, but not proved here. Exercise 13 requests a proof and suggests a particular approach. The inversion formula holds for example when f is a smooth function that tends to zero rapidly at infinity. As usual $\mathcal{S}$, in honor of Laurent Schwartz, denotes the space of such functions.

Theorem V.5.10. The Fourier transform is injective from $\mathcal{S}$ to $\mathcal{S}$ and the Fourier inversion formula holds:

$$f(x) = \left(\frac{1}{2\pi}\right)^{\frac{n}{2}} \int_{\mathbf{R}^n} e^{ix\cdot\xi} f^\wedge(\xi) d\xi.$$

Let α be a multi-index. For $f \in \mathcal{S}$ we apply the differential operator D^α to both sides of the inversion formula, and differentiate under the integral sign. This yields

$$(D^\alpha f)(x) = \left(\frac{1}{2\pi}\right)^{\frac{n}{2}} \int_{\mathbf{R}^n} (i\xi)^\alpha e^{ix\cdot\xi} f^\wedge(\xi) d\xi.$$

Thus we have the operator formula

$$D^\alpha = \mathcal{F}^{-1} M_{(i\xi)^\alpha} \mathcal{F},$$

which reveals the sense in which the Fourier transform *diagonalizes* differentiation. Furthermore, the operator formula makes sense if α is an n-tuple of real numbers, rather than an n-tuple of nonnegative integers. From the operator formula we obtain

$$D^\alpha D^\beta = D^{\alpha+\beta}.$$

More generally still we may replace the operation $M_{(i\xi)^\alpha}$ by M_p for more general functions p, depending on both x and ξ. The function p is called the *symbol* of the operator P, where P is defined by

$$P = \mathcal{F}^{-1} M_p \mathcal{F}.$$

In order to have a useful calculus of operators, one restricts the symbols appropriately. These considerations lead to the calculus of *pseudodifferential operators*.

One interesting aspect of the theory is the notion of the *order* of a pseudodifferential operator. A partial differential operator of order m is a pseudodifferential operator of order m. The fractional derivative operators D^α are of order $|\alpha| = \sum \alpha_j$. Operators of negative order are compact on $L^2(\mathbf{R}^n)$. Finally, the commutator $[A, B]$ of operators of orders a and b is of order at most $a+b-1$. In particular, the commutator of two pseudodifferential operators of order zero is compact.

Remark V.5.11. An important step in our work in Chapter VII is to show that a certain commutator $[P, M]$ is a compact operator on $L^2(\mathbf{B}_n)$. This result does not follow directly from the theory of pseudodifferential operators, because the operator P there is not a pseudodifferential operator. There is a more general class of operators, called *Fourier Integral Operators*, whose calculus does guarantee that this commutator is compact.

Exercise 11. Prove the composition formula $T_b T_a = T_{b+a}$ for the operators defined in Example V.5.9. Suggestion: Interchange the order of integration, change variables in the inner integral, and use the definition of the Euler beta function.

Exercise 12. Prove that the function $x \to e^{\frac{-x^2}{2}}$ is its own Fourier transform. Suggestion: Complete the square in the exponent, and use the calculus of residues to show that

$$\int_{\mathbf{R}} e^{\frac{-(t+is)^2}{2}}\, dt$$

is independent of s. (Here s and t are real.)

Exercise 13. Prove the Fourier inversion formula on $\mathcal{S}$. Suggestion: Write the right-hand side of the inversion formula as a double integral. Introduce a factor $e^{\frac{-\epsilon^2|\xi|^2}{2}}$ and use it to justify changing the order of integration. Use Exercise 12 and finally let ϵ tend to 0.

Exercise 14. Show that $\mathcal{F}^4 = I$. Then find all eigenfunctions of $\mathcal{F}$ on $L^2(\mathbf{R}^n)$.

CHAPTER VI

Positivity Conditions for Real-valued Functions

So far, most of our real-valued functions on $\mathbf{C}^n$ have been Hermitian forms such as $\langle Lz, z \rangle$. These are linear in z and in $\bar{z}$. When L is nonnegative definite, $\langle Lz, z \rangle = || \sqrt{L}\,(z) ||^2$; the squared norm makes it evident why the values are nonnegative. In this chapter we consider more general real-valued functions of several complex variables, and we apply some of the ideas from Hermitian linear algebra in this non-linear setting. Things become more interesting, even for real-valued polynomials.

Section 1 helps motivate the rest of the Chapter by considering positivity conditions for polynomials in one and several *real* variables. Section 2 returns us to material on real-valued functions of several complex variables. The rest of the Chapter studies reality and positivity conditions for polynomials.

VI.1 Real variables analogues

The main goal in this book is Theorem VII.1.1. This result has a much simpler analogue in the setting of several real variables. This simpler result was proved by Pólya around 1925, in conjunction with Artin's solution of Hilbert's seventeenth problem. Pólya's result is interesting even in one variable, where it had been noted earlier by Poincaré. This

175

section provides a self-contained discussion of the one-variable result. Pólya's Theorem (in several variables) will appear as Corollary VII.1.6. See [R] and [HLP] for related results and references; see [HLP] for an elementary proof of Pólya's Theorem.

First consider polynomials in one real variable x with real coefficients, and write $\mathbf{R}[x]$ to denote the set of such polynomials. Our emphasis on inequalities suggests two problems:

Problem VI.1.1. Characterize those $p \in \mathbf{R}[x]$ such that $p(x) > 0$ for all nonnegative x.

Problem VI.1.2. Characterize those $p \in \mathbf{R}[x]$ such that $p(x) \geq 0$ for all real x.

We say that the coefficients of a polynomial p of degree d are *all positive* if $p(x) = \sum_{j=0}^{d} c_j x^j$ and $c_j > 0$ for $0 \leq j \leq d$. We sometimes say instead p has *positive coefficients* with the same meaning. Note for example that the coefficients of the polynomial $1 + x^2$ are *not* all positive, as the coefficient c_1 vanishes.

If the coefficients of p are all positive, then $p(x) > 0$ for $x \geq 0$. The converse assertion is false; $x^2 - 2x + 2$ provides an easy counterexample. Proposition VI.1.3 states however that a polynomial in $\mathbf{R}[x]$ whose values are positive for all nonnegative x must be the *quotient* of polynomials whose coefficients are all positive. Conversely, the quotient $\frac{g}{h}$ of two such polynomials (whether or not it is a polynomial) has positive values at nonnegative x. Proposition VI.1.3 therefore solves Problem VI.1.1. Before stating and proving it we make three elementary observations.

First, let G_n denote the polynomial given by $G_n(x) = (1 + x)^n$. The coefficients of G_n are all positive, and thus $G_n(x)$ is positive for all nonnegative x. Second, the product of polynomials whose coefficients are all positive also has this property. Third, every polynomial in $\mathbf{R}[x]$ can be factored into a product of linear and quadratic factors in $\mathbf{R}[x]$.

Proposition VI.1.3. Suppose $p \in \mathbf{R}[x]$. Then $p(x) > 0$ for all non-negative x if and only if there is an integer N and a polynomial r with positive coefficients such that

$$(1 + x)^N p(x) = G_N(x)p(x) = r(x). \tag{1}$$

Proof. Suppose first that (1) holds for some N. Then $p(x) > 0$ for $x \geq 0$ because this is true for both G_N and r. The interest therefore lies in the converse assertion. We start with p and need to find N such that (1) holds.

Here is the strategy of the proof. First we factor p into a product of linear and quadratic polynomials:

$$p = \prod l_j \prod q_j.$$

Since $p(x)$ is positive for $x \geq 0$, we may assume each linear factor l_j has all positive coefficients. Suppose the conclusion of the Proposition holds for each quadratic factor q_j. Then, for each j, there is an integer N_j such that $G_{N_j} q_j = r_j$, where r_j has positive coefficients. Let $N = \sum N_j$. We then have

$$G_N \, p = \prod l_j \prod r_j = r$$

where r, defined by the formula, has all positive coefficients. The positivity follows because the product of polynomials with all positive coefficients also has all positive coefficients. Thus, to prove Proposition VI.1.3, it suffices to verify it for quadratic polynomials.

Suppose now that p is a quadratic polynomial, and $p(x) > 0$ when $x \geq 0$. The positivity for large x guarantees that the coefficient of x^2 is positive; after dividing by this coefficient we may assume $p(x) = x^2 + bx + c$. If $b \geq 0$, the conclusion is true with $N = 1$. If $b < 0$, we may first replace x by the positive number $\frac{-bx}{2}$, and then divide through by the positive number $\frac{b^2}{4}$. By doing so we reduce to the case where $p(x) = x^2 - 2x + 1 + m$, where $m > 0$, and m is the minimum value of p on all of $\mathbf{R}$.

We compute $r(x) = (1 + x)^n p(x)$ explicitly using the binomial theorem. We obtain

$$r(x) = x^{n+2} + (n-2)x^{n+1}$$

$$+ \sum_{k=2}^{n} C(n,k)x^k + (n(1+m) - 2)x + 1 + m \qquad (2)$$

where the coefficient $C(n,k)$ is given by

$$C(n,k) = \binom{n}{k-2} - 2\binom{n}{k-1} + (m+1)\binom{n}{k}$$

$$= \binom{n}{k}\left(1 + m + \frac{-2k}{n-k+1} + \frac{k(k-1)}{(n-k+2)(n-k+1)}\right).$$

$$(3)$$

We assume that $n \geq 3$, so the coefficients in (2) of x and x^{n+1} will be positive. It remains to study the coefficients $C(n,k)$. We work with $D(n,k)$, where

$$D(n,k) = \frac{C(n,k)}{\binom{n}{k}}$$

$$= m + 1 + \frac{-2k}{n-k+1} + \frac{k(k-1)}{(n-k+2)(n-k+1)}.$$

We replace k by αn for $0 \leq \alpha \leq 1$:

$$D(n,n\alpha) = m + 1 + \frac{-2\alpha}{1 - \alpha + \frac{1}{n}} + \frac{\alpha(\alpha - \frac{1}{n})}{(1 - \alpha + \frac{2}{n})(1 - \alpha + \frac{1}{n})}$$

$$= \frac{(m+1)(1 - \alpha + \frac{2}{n})(1 - \alpha + \frac{1}{n}) - 2\alpha(1 - \alpha + \frac{2}{n}) + \alpha(\alpha - \frac{1}{n})}{(1 - \alpha + \frac{2}{n})(1 - \alpha + \frac{1}{n})}.$$

$$(4)$$

Notice that formally letting n tend to infinity in (4) gives us $p(\frac{\alpha}{1-\alpha})$. We exploit this to finish the proof.

To prove the Proposition it suffices to show that there is a positive number ϵ and an integer n such that $D(n,k) \geq \epsilon$ for $2 \leq k \leq n$. It is easier to prove a stronger assertion. We will show that there is an n such that $D(n, n\alpha) \geq \epsilon$ for all $\alpha \in [0, 1]$.

For each n, the numerator in (4) is a quadratic polynomial in α. The sequence of these polynomials converges uniformly on $[0, 1]$ (by direct estimation) to the polynomial g given by

$$g(\alpha) = (m + 4)\alpha^2 - (4 + 2m)\alpha + (m + 1).$$

The minimum of g is easily seen to be the positive number $\frac{m}{m+4}$. By uniform convergence there is an n so that the numerator in (4) exceeds $\frac{m}{2(m+4)}$ for all $\alpha \in [0, 1]$. For this fixed n, the denominator in (4) is bounded above; therefore (4) is bounded below by a positive constant.

Thus we may choose n sufficiently large to ensure that each $C(n, k)$ is positive. Then r will have positive coefficients. We have verified the conclusion for quadratic polynomials, and therefore in general. $\qquad\square$

The proof of the positivity of $C(n, k)$ expresses this coefficient as a positive integer multiple of something asymptotically equal, as n tends to infinity, to a value of p. In other words, we can approximate the coefficients of $(1 + x)^n p(x)$ by binomial coefficients times values $p(\frac{k}{n-k})$ of p.

Proposition VI.1.3 suggests limit theorems from probability. Speaking in an imprecise manner, we can explain the result by saying that the finitely many dips in the graph of p get swamped by the rapid increase of G_n for large n. It is interesting that analogous results for smooth functions or convergent power series fail to hold.

We now say a few words about Problem VI.1.2. The answer is elegant; a polynomial in one real variable has only nonnegative values if and only if it is a sum of squares of polynomials. Artin's solution of Hilbert's seventeenth Problem shows that a polynomial in several real variables has only nonnegative values if and only if it is a sum of squares of *rational* functions. See Exercise 7 for an example of a polynomial whose values are nonnegative but which cannot be written as a sum of squares of *polynomials*.

The one-variable case has several elementary proofs. Perhaps the simplest proof uses *complex* numbers. We leave the one-variable case as an exercise with a useful hint.

Exercise 1. Suppose that $p \in \mathbf{R}[x]$ and that $p(x) \geq 0$ for all x. Show that there are an integer N and polynomials g_j for $j = 1, \ldots, N$ such that

$$p(x) = \sum_{j=1}^{N} g_j(x)^2.$$

Suggestion. Think of p as a polynomial in one complex variable z, and factor p into a product of linear factors over $\mathbf{C}$. Since $p(x)$ is real for x real, each factor $z - w$ must be accompanied by a factor $(z - \overline{w})$. With $w = a + ib$ and $z = x$ we have

$$(z - w)(z - \overline{w}) = (x - a)^2 + b^2.$$

Thus each of these quadratic factors of p is a sum of squares. Since the product of sums of squares is also a sum of squares, p itself is a sum of squares.

Exercise 2. Prove the arithmetic-geometric mean inequality

$$n \prod_{j=1}^{n} x_j \leq \sum_{j=1}^{n} x_j^n$$

for n nonnegative real variables. Suggestion 1: The proof for $n = 2$ is easy. Use induction to establish the result when n is a power of two. Then show that the truth of the inequality in n variables implies its truth in $n - 1$ variables. Suggestion 2: Use Lagrange multipliers to maximize the product of n positive numbers given that they sum to unity. Then use homogeneity and substitution to obtain the result.

Pólya gave a version of Proposition VI.1.3 in several variables. Let $\mathbf{R}[x_1, \ldots, x_n]$ denote the ring of polynomials with real coefficients in n real variables. We say that $p \in \mathbf{R}[x_1, \ldots, x_n]$ is homogeneous of degree m if $p(tx) = t^m p(x)$ for all $t \in \mathbf{R}$ and for all $x \in \mathbf{R}^n$. Then p

is homogeneous of degree m if and only if it contains no monomials of degree not equal to m. We say that a homogeneous polynomial has *positive coefficients* if the coefficient of each of its monomials is positive. We let $s(x)$ denote the polynomial given by the sum of the variables: $s(x) = x_1 + x_2 + \cdots + x_n$.

Pólya's theorem holds in the homogeneous case. By setting one of the variables equal to unity, we obtain an inhomogeneous version. Proposition VI.1.3 is the special case when $n = 2$ and one of the variables is equal to unity.

Theorem VI.1.4. (Pólya) Let p be a homogeneous polynomial in n real variables of degree m with real coefficients. Suppose that $p(x) > 0$ for all nonzero x in the set determined by $x_j \geq 0$ for all j. Then there is an integer d and a homogeneous polynomial q of degree $m + d$ with positive coefficients such that

$$s(x)^d p(x) = (x_1 + \cdots + x_n)^d p(x) = q(x). \tag{5}$$

Theorem VI.1.4 will be derived as a simple corollary of Theorem VII.1.1. We give here a few comments on a direct proof, which uses the same approach as the proof of Proposition VI.1.3, although one cannot simplify to the quadratic case at the start. We compute $s(x)^d p(x)$ using the multinomial theorem and see that the coefficients in the result are close to values of p; hence the coefficients are all positive when d is sufficiently large. It is easiest to write the details by introducing an extra variable. See [HLP]. The proof in [HLP] has a nice interpretation. One considers the extra variable as a *homotopy* between different bases of the vector space of homogeneous polynomials of a given degree.

Remark VI.1.5. Pólya's theorem glimpses, in a special case, one of the primary issues in this book. Let r be a polynomial function on $\mathbf{C}^n$ whose values are nonnegative. Is r the quotient of squared norms of a holomorphic polynomial mapping? The answer in general is no. Pólya's theorem covers the case when r is bihomogeneous (Definition VI.2.4), positive away from the origin, and depends only on the absolute values of the complex variables. In this case the answer is yes.

There is some interest in finding the minimum d that works in Proposition VI.1.3 and Pólya's theorem. See [R] for example. There is no bound on d that holds for all polynomials of a fixed degree (at least 2) in a fixed number of variables. This state of affairs applies even for quadratic polynomials in one variable.

Example VI.1.6. Put $p(x) = x^2 - 2x + 1.1$. Using Mathematica, for example, one can see that $(1 + x)^{39} p(x)$ has a negative coefficient. On the other hand $(1 + x)^{40} p(x)$ has all nonnegative coefficients; but the coefficient c_{21} of x^{21} vanishes identically:

$$c_{21} = \binom{40}{19} - 2\binom{40}{20} + 1.1\binom{40}{21} = \frac{40!}{19!20!}\left(\frac{2.1}{21} - \frac{2}{20}\right) = 0. \quad (6)$$

It follows easily that all coefficients are positive for the first time when the exponent N from Proposition VI.1.3 is 41.

More generally we let $p_\epsilon(x) = x^2 - 2x + 1 + \epsilon$. The minimum exponent that works in Proposition VI.1.3 for p_ϵ tends to infinity as ϵ tends to zero. To see this easily, note as in (6) that the coefficient c_{N+1} of $(1 + x)^{2N} p_\epsilon$ vanishes identically when $\epsilon = \frac{2}{N}$. Therefore the minimum exponent exceeds $\frac{2}{\epsilon}$.   □

It is possible to obtain estimates for the minimum exponent in terms of additional information on the values of p. The same situation applies in Theorem VII.1.1.

VI.2   Real-valued polynomials on $\mathbf{C}^n$

Our starting point is a polynomial function $r : \mathbf{C}^n \times \mathbf{C}^n \to \mathbf{C}$ such that $r(z, \bar{z})$ is real for all z. We sometimes consider the function $(z, w) \to r(z, \overline{w})$, and sometimes we consider the function $z \to r(z, \bar{z})$. In the second case we say that we are evaluating r *on the diagonal*.

Let $r : \mathbf{C}^n \times \mathbf{C}^n \to \mathbf{C}$ be a polynomial. Our first step is to determine when r is real on the diagonal. One characterization involves the *matrix of coefficients* $(c_{\alpha\beta})$ of r. The entries in this matrix depend on the ordering of the multi-indices and on the coordinates used in $\mathbf{C}^n$. We fix an ordering of the multi-indices. Crucial properties, such as whether $(c_{\alpha\beta})$ is Hermitian or whether it is nonnegative definite, are independent of the choice of coordinates. See Exercise 4.

Proposition VI.2.1. **(Reality)** Suppose that

$$r(z, \overline{w}) = \sum_{\alpha,\beta} c_{\alpha\beta} z^\alpha \overline{w}^\beta \tag{7}$$

is a polynomial function from $\mathbf{C}^n \times \mathbf{C}^n$ to $\mathbf{C}$. The following three statements are equivalent:

R1) The matrix of coefficients $(c_{\alpha\beta})$ is Hermitian.

R2) $r(z, \overline{w}) = \overline{r(w, \overline{z})}$ for all z and w.

R3) $r(z, \overline{z})$ is real for all z.

Proof. It is trivial that R2) implies R3). If R3) holds, then

$$\sum_{\alpha,\beta} c_{\alpha\beta} z^\alpha \overline{z}^\beta = \sum_{\alpha,\beta} \overline{c_{\alpha\beta}} \, \overline{z}^\alpha z^\beta = \sum_{\beta,\alpha} \overline{c_{\beta\alpha}} z^\alpha \overline{z}^\beta. \tag{8}$$

Equating coefficients in (8) shows that

$$c_{\alpha\beta} = \overline{c_{\beta\alpha}}$$

for all α, β; hence R1) holds. Thus R3) implies R1).

To see that R1) implies R2), write $\overline{r(w, \overline{z})}$ in terms of the coefficients, use the Hermitian symmetry, and interchange indices to obtain R2). $\qquad\Box$

This result is precisely analogous to Proposition IV.1.4. The reality conditions R1), R2), and R3) on the polynomial $r(z, \overline{z})$ are the same as the reality conditions on the Hermitian form $\langle Lz, z \rangle$. We could have proved Proposition VI.2.1 by deriving it from Proposition IV.1.4.

Since the matrix of coefficients of a real-valued polynomial is Hermitian, we obtain the following useful result.

Theorem VI.2.2. (Holomorphic decomposition) Suppose that

$$r(z, \overline{z}) = \sum_{\alpha, \beta} c_{\alpha\beta} z^{\alpha} \overline{z}^{\beta}$$

is a polynomial function from $\mathbf{C}^n \times \mathbf{C}^n$ to $\mathbf{R}$. Then there are holomorphic polynomial mappings f and g such that f and g take values in finite-dimensional spaces and

$$r(z, \overline{z}) = \| f(z) \|^2 - \| g(z) \|^2. \tag{9}$$

Proof. There are several proofs; one uses the spectral theorem. We give the most elementary proof, by providing formulas for f and g. Put

$$f_{\beta}(z) = \frac{1}{2} \left(z^{\beta} + \sum_{\alpha} c_{\alpha\beta} z^{\alpha} \right)$$

$$g_{\beta}(z) = \frac{1}{2} \left(z^{\beta} - \sum_{\alpha} c_{\alpha\beta} z^{\alpha} \right).$$

Since $r(z, \overline{z})$ is real, we have

$$\sum_{\beta} | f_{\beta}(z) |^2 - \sum_{\beta} | g_{\beta}(z) |^2 = \frac{1}{4} \sum_{\beta} 4 \operatorname{Re} \left(\overline{z}^{\beta} \sum_{\alpha} c_{\alpha\beta} z^{\alpha} \right)$$

$$= \sum_{\beta, \alpha} c_{\alpha\beta} z^{\alpha} \overline{z}^{\beta} = r(z, \overline{z}).$$

Since the sums are finite, we obtain the desired conclusion.   □

Exercise 3. Show that we may choose the polynomial mapping g in (9) to vanish identically if and only if the matrix $(c_{\alpha\beta})$ is nonnegative definite. Suggestion: Use Theorem IV.5.2.

Holomorphic decomposition enables us to determine for example when the zero-set of r is a complex variety, and more generally whether the zero-set of r contains any complex analytic varieties of positive dimension. See [D] for these applications.

We make some simple observations about Theorem VI.2.2. First, the polynomials f_β and g_β arising in it may be chosen of degree at most the maximum degree in z of r. Second, the decomposition is not unique. We will discuss how decompositions are related after proving Theorem VI.2.3.

Next we introduce two useful operations on polynomial mappings. Let g, h be (holomorphic) polynomial mappings with domain $\mathbf{C}^n$ and targets $\mathbf{C}^m$ and $\mathbf{C}^k$. Their *orthogonal sum* $g \oplus h$ is the polynomial mapping with target $\mathbf{C}^{m+k}$ defined by

$$(g \oplus h)(z) = (g(z), h(z)).$$

Note that

$$\| (g \oplus h)(z) \|^2 = \| g(z) \|^2 + \| h(z) \|^2. \tag{10}$$

Their *tensor product* $g \otimes h$ is the polynomial mapping to $\mathbf{C}^{mk}$ whose components are all possible products $g_a h_b$ of the components of g and h in some determined order. Although the definition of $g \otimes h$ depends on the order, the squared norm does not. For each z,

$$\| (g \otimes h)(z) \|^2 = \| g(z) \|^2 \| h(z) \|^2. \tag{11}$$

A special case of (11) will play an important role. First consider the identity mapping $z \to z$ on $\mathbf{C}^n$. We may form its m-fold tensor product

$$z^{\otimes m} = z \otimes \cdots \otimes z$$

and use (11) to obtain

$$\| z^{\otimes m} \|^2 = \| z \|^{2m}.$$

In $z^{\otimes m}$ the various monomials are listed with repetitions. It is convenient to introduce also the related mapping H_m, where the components are linearly independent. Fix some ordering of the multi-indices of order m. Let $\binom{m}{\alpha}$ denote the multinomial coefficient $\frac{m!}{\prod \alpha_j!}$. We define H_m, where $|\alpha| = m$, by

$$H_m(z) = \left(\ldots, \sqrt{\binom{m}{\alpha}} z^\alpha, \ldots \right). \tag{12}$$

It follows from the multinomial theorem that

$$\| H_m(z) \|^2 = \| z \|^{2m} = \| z^{\otimes m} \|^2.$$

Note that H_m maps into a space of smaller dimension than $z^{\otimes m}$ does. The equality of their squared norms implies that the mappings H_m and $z^{\otimes m}$ are related via a linear transformation L. It follows from our next result that $H_m(z) = P(U(z^{\otimes m}))$, where U is unitary and P is a projection. See Exercise 5.

The next result has many applications (see [D]).

Theorem VI.2.3. Let A and B be holomorphic polynomial mappings from $\mathbf{C}^n$ to $\mathbf{C}^N$, and suppose $\| A(z) \|^2 = \| B(z) \|^2$ for all z. Then there is a unitary mapping U on $\mathbf{C}^N$ such that, for all z,

$$A(z) = U B(z).$$

Proof. We assume that $A(z) = \sum A_\alpha z^\alpha$ and $B(z) = \sum B_\alpha z^\alpha$ where the coefficients are elements of $\mathbf{C}^N$. The condition that $\| A(z) \|^2 = \| B(z) \|^2$ becomes, upon equating coefficients, that, for all α and β:

$$\langle A_\alpha, A_\beta \rangle = \langle B_\alpha, B_\beta \rangle. \tag{13}$$

Consider a maximal linearly independent set of the B_α, and define U on these by

$$U(B_\alpha) = A_\alpha.$$

If B_β is a linear combination of these B_α, then the definition of $U(B_\beta)$ is forced by linearity. On the other hand, the compatibility conditions

(13) show that U is a well-defined linear transformation on the span of the B_β, and that

$$\langle U B_\alpha, U B_\beta \rangle = \langle A_\alpha, A_\beta \rangle = \langle B_\alpha, B_\beta \rangle.$$

Therefore U preserves all inner products, and hence is the restriction of a unitary transformation on $\mathbf{C}^N$ to the span of the B_β. $\square$

Note that Theorem VI.2.3 has an obvious converse. When $A(z) = U B(z)$ for some unitary U, we have $\| A(z) \|^2 = \| B(z) \|^2$. We may use Theorem VI.2.3 to analyze the nonuniqueness in the choice of f and g in Theorem VI.2.2. The statement

$$\| f(z) \|^2 - \| g(z) \|^2 = \| F(z) \|^2 - \| G(z) \|^2$$

is equivalent to

$$\| f(z) \|^2 + \| G(z) \|^2 = \| g(z) \|^2 + \| F(z) \|^2.$$

By including component functions vanishing identically, we may assume that there are K functions on each side of this equality. Theorem VI.2.3 then guarantees the existence of a unitary mapping U on $\mathbf{C}^K$ with $f \oplus G = U(g \oplus F)$.

Exercise 4. Suppose $r : \mathbf{C}^n \times \mathbf{C}^n \to \mathbf{C}$ is a polynomial in z and $\bar{z}$. How does a linear change of coordinates alter the matrix of coefficients? Prove that Hermitian symmetry and nonnegative definiteness are preserved by such a change of coordinates.

Exercise 5. Let f and g be holomorphic polynomial mappings with values in finite-dimensional complex Euclidean spaces of different dimensions. Suppose that $\| f \|^2 = \| g \|^2$. How does Theorem VI.2.3 generalize?

We next introduce a natural homogeneity condition for polynomials.

Definition VI.2.4. **(Bihomogeneous polynomial)** A polynomial function

$$r(z, \overline{z}) = \sum_{|\alpha| = |\beta| = m} c_{\alpha\beta} z^{\alpha} \overline{z}^{\beta}$$

is called a bihomogeneous polynomial of degree $2m$.

A bihomogeneous polynomial r is homogeneous of degree m in z, and of degree m in $\overline{z}$. Thus, for all $\lambda \in \mathbf{C}$, we have $r(\lambda z, \overline{\lambda z}) = |\lambda|^{2m} r(z, \overline{z})$. This condition provides us with a more general definition.

Definition VI.2.5. **(Bihomogeneous function)** Let r be a continuous complex-valued function on $\mathbf{C}^n$, and suppose m is an arbitrary positive number. We say that r is *bihomogeneous* of degree $2m$ if, for all $\lambda \in \mathbf{C}$ and all $z \in \mathbf{C}^n$,

$$r(\lambda z) = |\lambda|^{2m} r(z). \tag{14}$$

In Definition VI.2.5 we insist that (14) hold for all complex numbers λ. The (real) homogeneous function $z \rightarrow z^2 + \overline{z}^2$ satisfies (14) with $m = 1$ for all *real* λ, but not for all *complex* λ. It is therefore not bihomogeneous.

Many bihomogeneous functions are not polynomials. One simple example is a fractional power of a bihomogeneous polynomial. Another class of examples arises from considering quotients of bihomogeneous polynomials, where the denominator vanishes only at the origin, and the numerator vanishes to higher order there.

We now begin to study bihomogeneous polynomials. By Proposition VI.2.1, a bihomogeneous polynomial is real-valued if and only if its matrix of coefficients $c_{\alpha\beta}$ is Hermitian. Example VI.3.1 from the next section, when $-2 < c < 0$, gives a bihomogeneous polynomial whose values away from the origin are positive, but for which the matrix of coefficients has a negative eigenvalue. Our next goal is to develop techniques for clarifying this situation.

The number of independent monomials of degree m in n variables is the binomial coefficient

$$d(n, m) = \binom{n + m - 1}{m}.$$

Consider complex Euclidean spaces of dimensions n and N, where $N = d(n, m)$. There is a mapping $\zeta : \mathbf{C}^n \to \mathbf{C}^N$, whose α-th coordinate is given by

$$\zeta_\alpha(z) = z^\alpha.$$

This mapping is called the *Veronese mapping* of degree m, and it suggests an interesting point of view on bihomogeneous polynomials.

Let r be a bihomogeneous polynomial on $\mathbf{C}^n$ of degree $2m$ with matrix of coefficients C. We may consider $C = (c_{\alpha\beta})$ to be the matrix of a Hermitian form on $\mathbf{C}^N$, where $N = d(n, m)$. We have, with $\zeta = \zeta(z)$,

$$\sum_{\alpha,\beta} c_{\alpha\beta} \zeta_\alpha \overline{\zeta}_\beta = \sum_{\alpha,\beta} c_{\alpha\beta} z^\alpha \overline{z}^\beta = r(z, \overline{z}). \tag{15}$$

Thus the polynomial r can be considered as the pullback to $\mathbf{C}^n$ via the Veronese mapping of the Hermitian form $\langle C\zeta, \zeta \rangle$ on $\mathbf{C}^N$. We write (15) as

$$r(z, \overline{z}) = \langle C\zeta(z), \zeta(z) \rangle.$$

Our proof of Theorem VI.2.2 was direct. It is convenient, when r is bihomogeneous, to apply the Hermitian linear algebra from Chapter IV to give another proof. This proof also works for arbitrary real-valued polynomials.

Theorem VI.2.6. Let r be a bihomogeneous polynomial of degree $2m$ on $\mathbf{C}^n$. There are holomorphic homogeneous polynomials

$f_1, \ldots, f_k, g_1, \ldots, g_l$ such that

$$r(z, \overline{z}) = \sum_{a=1}^{k} |f_a(z)|^2 - \sum_{b=1}^{l} |g_b(z)|^2$$
$$= \| f(z) \|^2 - \| g(z) \|^2.$$

Proof. Let $N = d(m, n)$. The matrix of coefficients of r defines a Hermitian operator on $\mathbf{C}^N$. By the spectral theorem (Theorem IV.4.2), there is a basis $\{v_j\}$ of eigenvectors with corresponding real eigenvalues λ_j for which we can write

$$\sum_{\alpha, \beta} c_{\alpha\beta} \zeta_\alpha \overline{\zeta}_\beta = \sum_{j=1}^{N} \lambda_j |\langle \zeta, v_j \rangle|^2$$

and hence

$$r(z, \overline{z}) = \sum_{j=1}^{N} \lambda_j |\langle \zeta(z), v_j \rangle|^2 = \sum_{j=1}^{N} \pm |\langle \zeta(z), \sqrt{|\lambda_j|} \, v_j \rangle|^2.$$

Since $\langle \zeta(z), v_j \rangle$ is a homogeneous polynomial in z of degree m, the desired conclusion follows. To summarize, we pull back via the Veronese mapping $\zeta_\alpha = z^\alpha$. $\qquad\square$

Corollary VI.2.7. Suppose that r is a bihomogeneous polynomial and that its matrix of coefficients is positive definite. Then

$$r(z, \overline{z}) = \sum_{a=1}^{d(m,n)} |f_a(z)|^2 = \| f(z) \|^2$$

where the polynomials f_a form a basis for the vector space of holomorphic homogeneous polynomials of degree m.

Corollary VI.2.8. Suppose that r is a bihomogeneous polynomial, that its matrix of coefficients is nonnegative definite, and that r is not identically zero. Then there is a k with $1 \leq k \leq d(n, m)$ and linearly independent homogeneous polynomials f_a such that

$$r(z, \overline{z}) = \sum_{a=1}^{k} | f_a(z) |^2 = \| f(z) \|^2.$$

Next we show how to transform a polynomial into an operator on $A^2(\mathbf{B}_n)$. Given a polynomial $r(z, \overline{z})$ we define a linear transformation $L : A^2(\mathbf{B}_n) \to A^2(\mathbf{B}_n)$ by using $r(z, \overline{\zeta})$ as an integral kernel. We apply these ideas in Chapter VII.

Definition VI.2.9. Given a polynomial $r : \mathbf{C}^n \times \mathbf{C}^n \to \mathbf{C}$, we define a linear transformation $L = L_r$ from $A^2(\mathbf{B}_n)$ to itself by

$$(Lh)(z) = \int_{\mathbf{B}_n} r(z, \overline{\zeta}) h(\zeta) \, dV(\zeta).$$

To understand L we introduce convenient notation. We write Φ_α for the (not normalized) monomial function $z \to z^\alpha$. Then we have

$$Lh = \sum_{\alpha, \beta} c_{\alpha\beta} \Phi_\alpha \langle h, \Phi_\beta \rangle. \tag{16}$$

In other words, L replaces $\overline{z}^\beta$ with the operation of taking the inner product with z^β. As a consequence, we have

$$\langle Lh, g \rangle = \sum_{\alpha, \beta} c_{\alpha\beta} \langle h, \Phi_\beta \rangle \overline{\langle g, \Phi_\alpha \rangle}. \tag{17}$$

Recall that $A^2(\mathbf{B}_n)$ is the orthogonal sum of the subspaces V_k generated by the monomials of degree k and that all pairs of distinct monomials are orthogonal. Thus we can understand these operators by understanding the bihomogeneous case.

Proposition VI.2.10. Let r be a bihomogeneous polynomial of degree $2m$. Let L denote the linear transformation on $A^2(\mathbf{B}_n)$ from Definition VI.2.9. The following four statements are true:

1) The restriction of L to V_k is the zero operator unless $k = m$.

2) L maps V_m into itself.

3) $L = L^*$ if and only if r is real on the diagonal, if and only if $r(z, \overline{w}) = \overline{r(w, \overline{z})}$.

4) The restriction of L to V_m is positive definite if and only if the matrix of coefficients of r is positive definite.

Proof. It follows directly from (16) and the orthogonality of monomials of different degrees that 1) and 2) hold. To prove 3) we use (17) and then interchange indices:

$$\langle Lh, g \rangle - \langle h, Lg \rangle = \langle Lh, g \rangle - \overline{\langle Lg, h \rangle}$$
$$= \sum (c_{\alpha\beta} - \overline{c_{\beta\alpha}}) \langle h, \Phi_\beta \rangle \overline{\langle g, \Phi_\alpha \rangle}. \qquad (18)$$

It follows that $L = L^*$ if and only if the matrix $C = (c_{\alpha\beta})$ is Hermitian symmetric. By Proposition VI.2.1, C is Hermitian symmetric if and only if the other two statements in 3) hold.

To prove 4), we note from (17) that

$$\langle Lh, h \rangle = \sum c_{\alpha\beta} \langle h, \Phi_\beta \rangle \overline{\langle h, \Phi_\alpha \rangle}.$$

Since the monomials of degree m form a basis for V_m, we see that $\langle Lh, h \rangle \geq c \|h\|^2$ if and only if the conjugated matrix $(\overline{c_{\alpha\beta}})$ is positive definite. This holds if and only if $(c_{\alpha\beta})$ is positive definite. $\qquad \square$

Remark VI.2.11. Notice how the matrix of coefficients gets conjugated when we pass from r to L_r. Since this matrix is Hermitian, taking conjugates is equivalent to taking transposes (without taking conjugates). This apparently strange phenomenon is a manifestation of duality.

Proposition VI.2.10 allows us to view a real-valued bihomogeneous polynomial of degree $2m$ as a Hermitian operator from V_m to itself.

VI.3 Squared norms and quotients of squared norms

In Proposition VI.2.1 we noted three simple equivalent conditions for the *reality* of a polynomial r on $\mathbf{C}^n$. In particular, we can determine whether r is real-valued by looking at the matrix of coefficients. Things are more subtle when we consider the *positivity* of r. When the matrix of coefficients is nonnegative definite, it is valid to conclude that r has only nonnegative values. The converse is false. The next examples show that r can have only nonnegative values even when the matrix has a negative eigenvalue.

Example VI.3.1. Put $r(z, \bar{z}) = |z_1|^4 + c|z_1 z_2|^2 + |z_2|^4$. The matrix of coefficients is diagonal, with eigenvalues $1, c, 1$. The condition for nonnegativity of the matrix is thus $c \geq 0$. The condition for nonnegativity of the function values is $c \geq -2$, as is evident by completing the square.

Example VI.3.2. Consider the polynomial p defined by

$$p(z, \bar{z}) = |z_1|^6 + a|z_1|^4|z_2|^2 + b|z_1|^2|z_2|^4 + |z_2|^6$$

where a and b are real numbers. The matrix is diagonal with eigenvalues $1, a, b, 1$. The condition for nonnegativity of this polynomial (Exercise 6) allows a or b to be arbitrarily negative as long as the other is sufficiently positive.

Exercise 6. Determine the condition on a and b so that the polynomial p in Example VI.3.2 is positive away from the origin.

In this section we use the term *squared norm* to mean a finite sum of squared absolute values of holomorphic polynomial mappings. Such functions achieve only nonnegative values; we will need to study more general conditions guaranteeing nonnegativity. One useful larger class of nonnegative functions consists of those polynomials that are quotients of squared norms of holomorphic polynomial mappings. We call such polynomial functions *quotients of squared norms.*

Being a quotient of squared norms places some interesting restrictions on a polynomial r. To find these conditions we follow a general principle. To obtain information about a function of several complex variables, we study its restriction to all holomorphic curves in its domain. Combining this principle with our interest in polynomials suggests the following definition.

Definition VI.3.3. A *curve* in $\mathbf{C}^n$ is a mapping $t \to z(t)$ such that:
  There is a holomorphic function $q : \mathbf{C} \to \mathbf{C}$, not identically zero, and a holomorphic mapping $p : \mathbf{C} \to \mathbf{C}^n$ with

$$z(t) = \frac{p(t)}{q(t)}.$$

When q is a constant we call such a curve a *holomorphic curve*. In general the denominator q may vanish at some points, so a curve defines a mapping into $\mathbf{C}^n$ only away from the zero-set of q.

Suppose that r is a real-analytic function that can be written as the quotient of squared norms of holomorphic polynomial mappings. Let $t \to z(t)$ be a curve in $\mathbf{C}^n$. We write z^*r for the restriction, or pullback, of r to z. This is the function $t \to r(z(t), \overline{z(t)})$.

The pullback z^*r to a holomorphic curve of a quotient of squared norms is far from arbitrary. Substitution and simplification give

$$z^*r(t, \bar{t}) = \frac{\| f(z(t)) \|^2}{\| g(z(t)) \|^2} = \frac{\| f((\frac{p}{q})(t)) \|^2}{\| g((\frac{p}{q})(t)) \|^2} = \frac{\| A(t) \|^2}{\| B(t) \|^2},$$

where A and B are holomorphic mappings obtained by clearing denominators. Here we are using the information that f and g are polynomials. Suppose that $A(t)$ is not identically zero. We obtain

$$z^*r(t, \bar{t}) = \frac{\| at^m + \cdots \|^2}{\| bt^k + \cdots \|^2} = c|t|^{2(m-k)} + \cdots \qquad (19)$$

where the omitted terms are of higher degree, and (the vectors) a and b are assumed not zero. Notice that the dots inside the squared norms involve higher powers of t, whereas those on the right-hand side of (19) involve terms in t and $\bar{t}$.

The exponent $2(m-k)$ can be negative, but it is always an even integer. From (19) we see that there are no other terms of order $2(m-k)$; the part of the series for z^*r of lowest order must be independent of the *argument* of t. Thus, if $t = |t|e^{i\theta}$, the lowest order part is independent of θ. The author named this necessary condition for being a quotient of squared norms the *jet pullback property*. In differential topology *jets* are essentially the same as Taylor polynomials.

Definition VI.3.4. Let r be a real-analytic real-valued function defined on $\mathbf{C}^n$. We say that r satisfies the *jet pullback property* if, for each curve z, either 1) or 2) holds:

1) The pullback z^*r vanishes identically.

2) There is a (possibly negative) even integer $2s$ such that the following limit exists and is not zero:

$$\lim_{t \to 0} |t|^{2s} r(z(t), \overline{z(t)}). \tag{20}$$

In case z^*r is smooth at $t = 0$, and 1) does not hold, then we choose $2s$ in (20) to be minus the order $2a$ of vanishing of z^*r. The existence of the limit in (20) is equivalent to

$$z^*r(t, \bar{t}) = c|t|^{2a} + \cdots,$$

where the omitted terms are of order greater than $2a$. The presence of any terms of order $2a$ other than $c|t|^{2a}$ prevents the limit from existing.

Example VI.3.5. Let $n = 1$ and put

$$r(z, \bar{z}) = 1 + z^2 + \bar{z}^2 + 2|z|^2 = 1 + 4x^2.$$

Then $r(z, \bar{z}) \geq 1$ for all z. Consider the curve defined by $z(t) = \frac{1}{t}$.

Since

$$z^*r(t, \bar{t}) = |t|^{-4}(2|t|^2 + t^2 + \bar{t}^2 + |t|^4),$$

the jet pullback property fails. Hence r *cannot* be written as the quotient of squared norms. It follows that x^2 is not a quotient of squared norms either.

Example VI.3.6. Put $r(z, \bar{z}) = (|z_1 z_2|^2 - |z_3|^4)^2 + |z_1|^8$. Then r is a bihomogeneous polynomial of degree 8 in three variables such that:

1) r assumes only nonnegative values,

2) the zero-set of r is a complex line,

3) r is not a quotient of squared norms, because the jet pullback property fails.

Statement 1) is evident. The zero-set of r is the complex line given by $z_1 = z_3 = 0$ and z_2 is arbitrary, so statement 2) holds as well. To see that the jet pullback property fails, we choose the curve $t \to (t^2, 1 + t, t)$. Then

$$z^*r(t, \bar{t}) = t^4 \bar{t}^6 + 2|t|^{10} + t^6 \bar{t}^4 + \cdots$$

and the lowest order terms are not acceptable.

Exercise 7. Define a homogeneous polynomial p on $\mathbf{R}^3$ by

$$p(x) = x_1^4 x_2^2 + x_1^2 x_2^4 + x_3^6 - 3x_1^2 x_2^2 x_3^2.$$

Prove that $p(x) \geq 0$ for all x. Suggestion: Apply the arithmetic-geometric mean inequality (Exercise 2). Prove that p is not a sum of squares of polynomials.

Exercise 8. Define p by

$$p(z, \bar{z}) = |z_1|^4 |z_2|^2 + |z_1|^2 |z_2|^4 + |z_3|^6 - 3|z_1 z_2 z_3|^2. \quad (21)$$

By Exercise 7, $p(z, \overline{z}) \geq 0$ for all z. Find a curve $t \to z(t)$ such that $z^* p$ violates the jet pullback property.

VI.4 Plurisubharmonic functions

Plurisubharmonic functions are the complex variable analogues of convex functions. They play a vital role in complex analysis and complex differential geometry. We cannot cover the subject in depth, so we provide a simplified treatment of the basic issues. Some examples help set the stage for the next section, where we introduce many *positivity conditions* for real-valued polynomials of several complex variables.

We recall that a function of one real variable is called *convex* if the following geometric statement holds. For each pair of points x, y with $x \leq y$, the graph of f on the interval $[x, y]$ lies on or below the graph of the line segment joining $(x, f(x))$ and $(y, f(y))$. For real numbers a and b, and $t \in [0, 1]$, the expression $(1 - t)a + tb$ is called a convex combination of a and b. It parametrizes the line segment from a to b as t ranges from 0 to 1. The definition of convexity is thus evidently equivalent to the validity of the following inequality for all x, y in the domain of f, and for all $t \in [0, 1]$:

$$f((1 - t)x + ty) \leq (1 - t)f(x) + tf(y). \qquad (22)$$

The convexity inequality (22) provides also the definition of a convex function of several (or even infinitely many) variables. Convex functions are not necessarily differentiable; on the other hand a differentiable function of one variable is convex if and only if its derivative is nondecreasing. A twice differentiable function of one variable is convex if and only if its second derivative is nonnegative. This result generalizes to several variables; a twice differentiable function of several real variables is convex if and only if its second derivative matrix is nonnegative definite.

Exercise 9. Show that (22) implies that the average value of f on the interval $[x, y]$ is at most the average of the values of f at the endpoints.

Definition VI.4.1. Let r be a smooth function of n complex variables. Its *complex Hessian* is the n-by-n matrix

$$H(r) = \begin{pmatrix} r_{z_1\bar{z}_1} & r_{z_1\bar{z}_2} & \cdots & r_{z_1\bar{z}_n} \\ r_{z_2\bar{z}_1} & r_{z_2\bar{z}_2} & \cdots & r_{z_2\bar{z}_n} \\ \cdots & \cdots & \cdots & \cdots \\ r_{z_n\bar{z}_1} & \cdots & \cdots & r_{z_n\bar{z}_n} \end{pmatrix}.$$

We will use without proof the second-derivative test: If a smooth real-valued function r has a local minimum at p, then $H(r)$ is nonnegative definite at p.

Our preliminary discussion suggests the definition of subharmonic and plurisubharmonic for smooth functions.

Definition VI.4.2. (subharmonic and plurisubharmonic functions)
Let D be an open subset of $\mathbf{C}$. A smooth function $r : D \to \mathbf{R}$ is called *subharmonic* if its Laplacian $r_{z\bar{z}}$ is a nonnegative function. Let Ω be an open subset of $\mathbf{C}^n$. A smooth function $r : \Omega \to \mathbf{R}$ is called *plurisubharmonic* if its complex Hessian $H(r)$ is nonnegative definite at each point of Ω. The function r is called *strongly plurisubharmonic* on Ω if $H(r)$ is positive definite at each point of Ω.

Thus r is plurisubharmonic on $\mathbf{C}^n$ if, for all $z, a \in \mathbf{C}^n$, we have

$$H(r)(a, a) = \sum_{j,k=1}^{n} r_{z_j\bar{z}_k}(z)a_j\bar{a}_k \geq 0.$$

Suppose that L is a Hermitian matrix, such as the Hessian $H(r)$ at a fixed point. Henceforth we will often write $L \geq 0$ or $H(r) \geq 0$ instead of saying that L or $H(r)$ is nonnegative definite.

Remark VI.4.3. Definition VI.4.2 considers only smooth functions. A more general definition uses a mean-value inequality, analogous to (5) from Chapter III for harmonic functions and to (22) for convex functions. A continuous function r is subharmonic if, for each ball in its domain, the value $r(z)$ at the center of the ball is at most the average value of r over that ball. A continuous function is plurisubharmonic if its restriction to every complex line is subharmonic. For smooth functions, this property is equivalent to the nonnegativity of the complex Hessian.

The word "plurisubharmonic" might sound a bit strange, because the prefixes *pluri* and *sub* seem to contradict each other. They do not; *pluri* means "more" as in "more than one variable," whereas *sub* means "less" as in "less than or equal to," in reference to the mean-value inequality from Remark VI.4.3. Thus "plurisubharmonic" does not mean "more or less harmonic."

Remark VI.4.4. Nonnegativity of the Hessian (in both the real and complex cases) plays a crucial role in partial differential equations and harmonic analysis. The notion applies even for continuous functions. Mathematicians have developed *Distribution Theory*, where one considers *weak derivatives*. The derivative of a (possibly nondifferentiable) function is defined to be a continuous linear functional on an appropriate space of functions. Once this notion has been defined, one can characterize plurisubharmonic functions as those whose complex Hessian matrix, in the sense of distributions, is nonnegative. See [Do], [H], [Kr], [S], or [Ta] for information about these topics.

Exercise 10. Let $n = 1$ and put $r_a(z) = z^2 + \bar{z}^2 + a|z|^2$ for a real parameter a. Show that r is subharmonic if and only if $a \geq 0$, and show that r_a is convex (as a function of two real variables) if and only if $a \geq 2$.

Exercise 11. Suppose that f is twice differentiable on an open set in $\mathbf{R}^2$ and convex there. Show that f, considered as a function of the complex variable z, is subharmonic. Generalize to higher dimensions.

Readers familiar with real variable theory know that the class of positive functions whose logarithms are convex plays a key role. An analogous definition is useful in complex variable theory.

Definition VI.4.5. Let r be a nonnegative smooth function. Then r is called *logarithmically plurisubharmonic* if the matrix $M(r)$ is nonnegative definite at each point:

$$M(r) = (r\, r_{z_j \bar{z}_k} - r_{z_j} r_{\bar{z}_k}) \geq 0.$$

It is convenient to introduce some simpler notation. We write $\langle \partial r, \overline{a} \rangle$ instead of $\sum r_{z_j} a_j$, and we write $|\,\partial r\,|^2$ for the Hermitian form defined by $|\,\partial r\,|^2(a, a) = |\,\langle \partial r, \overline{a} \rangle\,|^2$:

$$|\,\partial r\,|^2(a, a) = \sum r_{z_j} r_{\bar{z}_k} a_j \overline{a}_k$$

The matrix of this Hermitian form has rank one where ∂r does not vanish. As the notation suggests, it is always nonnegative definite.

These conventions enable us to write $M(r)$ in simpler notation:

$$M(r) = r H(r) - |\,\partial r\,|^2.$$

This formula gives one of several ways to see the simple result that logarithmic plurisubharmonicity implies plurisubharmonicity. Since r is nonnegative, the nonnegativity of $M(r)$ implies the nonnegativity of $H(r)$ wherever r doesn't vanish. On the other hand, where r vanishes, $H(r)$ is necessarily nonnegative because r has a minimum there. Another approach is to consider more generally the effect on the Hessian of composing a function with a convex increasing function of one real variable.

Let r be positive. Computing the Hessian of $\log(r)$ using the chain rule gives

$$r^2 H(\log(r)) = M(r).$$

When r is positive it is evident that r is logarithmically plurisubharmonic ($M(r) \geq 0$) if and only if $\log(r)$ is plurisubharmonic ($H(\log(r)) \geq 0$). In the general theory of plurisubharmonic functions one allows functions to achieve the value $-\infty$. This characterization then holds for *nonnegative* functions r. Hence, even when r takes on the value 0, we say "$\log(r)$ is plurisubharmonic" instead of saying "r is logarithmically plurisubharmonic."

We indicate why it is important in complex analysis to allow the value $-\infty$. If f is holomorphic, then $\log(|f|)$ is plurisubharmonic where f doesn't vanish. To study the zeroes of f using the theory of plurisubharmonic functions, we must confront the value $-\infty$.

We give several examples before we return to the general discussion.

Example VI.4.6. The Hermitian form $\langle Az, z \rangle$ defines a plurisubharmonic function if and only if A is nonnegative definite, and hence if and only if $\langle Az, z \rangle = \|Tz\|^2$ for some linear transformation T.

Example VI.4.7. Suppose $n = 1$. Put $r(z, \overline{z}) = |z|^2 + c(z^2 + \overline{z}^2)$. It is evident that r is subharmonic for all c and that its values are nonnegative for $|c| \leq \frac{1}{2}$. A simple computation shows that $\log(r)$ is subharmonic only when $c = 0$. (Recall that this r is homogeneous but not bihomogeneous.)

Example VI.4.8. Suppose $n = 1$. Put $r(z, \overline{z}) = (|z|^2 + 1)^4 - a|z|^4$. This r is the restriction to the plane given by $z_2 = 1$ of the function in Proposition VI.5.6. Here $\log(r)$ is subharmonic if and only if $a \leq 12$, while r is subharmonic if and only if

$$a \leq \frac{3}{32}(69 + 11\sqrt{33}) \approx 12.39.$$

Proof. First we observe, for $a \leq 6$, that r is a squared norm, and hence both r and $\log(r)$ are subharmonic. In the calculations we therefore assume that $a > 6$. The Laplacian of r is easily found:

$$r_{z\overline{z}} = 16|z|^6 + 36|z|^4 + (24 - 4a)|z|^2 + 4.$$

Write $t = |z|^2$. We see that r is subharmonic if and only if

$$u(t) = 4t^3 + 9t^2 + (6 - a)t + 1 \geq 0$$

holds for all nonnegative t. The function u has one negative root, it is positive at $t = 0$, decreasing at $t = 0$, and is positive for all large t. Thus it will be nonnegative for all nonnegative t if and only if its min-

imum value (for $t \geq 0$) is nonnegative. Thus we can find the largest a
for which u is nonnegative by choosing a so that u has a double root
for some positive t_0. We first solve the equation $u'(t_0) = 0$ for t_0. We
find that the nonnegative solution is given by

$$t_0 = \frac{-18 + \sqrt{36 + 48a}}{24}.$$

We set $u(t_0) = 0$ and solve for a. We obtain the value claimed.

Next we indicate the proof that $\log(r)$ is subharmonic if and only
if $a \leq 12$. We compute the expression $D = r r_{z\bar{z}} - |r_z|^2$ and write it in
terms of t. We obtain

$$D(t) = 4(1+t)^2 \left(1 + (4-a)t + (6+a)t^2 + (4-a)t^3 + t^4 \right).$$

We do not alter the positivity if we divide by $4(1+t)^2$. Therefore we
analyze

$$\begin{aligned} d_a(t) &= 1 + (4-a)t + (6+a)t^2 + (4-a)t^3 + t^4 \\ &= (1+t)^4 + a(-t + t^2 - t^3). \end{aligned}$$

This expression is linear in a, and the coefficient of a is nonpositive
when $t \geq 0$. Therefore if $d_a(t)$ is nonnegative for $t \geq 0$, then the same
holds for all smaller values of a. When $a = 12$, we see that $d_a(t)$ is a
perfect square:

$$d_{(12)}(t) = (1 - 4t + t^2)^2.$$

Note that this polynomial has zeroes for positive t. When $a > 12$,
we see that $d_a(t)$ has negative values. Therefore the condition for the
subharmonicity of $\log(r)$ is $a \leq 12$.   □

Our next propositions reveal some connections between plurisubharmonicity and logarithmic plurisubharmonicity.

Proposition VI.4.9. Suppose that r is a smooth positive function.
Then $\log(r)$ is plurisubharmonic if and only if $|e^{a(z)}|^2 r(z)$ defines a

plurisubharmonic function for each linear function a. (Here $a(z) = \sum a_k z_k$.)

Proof. Suppose first that $\log(r)$ is plurisubharmonic. Since $z \to 2\operatorname{Re}(a(z))$ is harmonic, we see that

$$\log(r) + 2\operatorname{Re}(a)$$

is also plurisubharmonic. Since exponentiation is a convex increasing function on **R**, the exponential of a plurisubharmonic function is also plurisubharmonic (Exercise 13). Exponentiating shows that $|e^a|^2 r$ is a plurisubharmonic function.

To show the converse assertion we first define u by

$$u(z) = |e^{a(z)}|^2 r(z).$$

We will compute the Hessian of u at an arbitrary point, and then choose a intelligently. By assumption u is plurisubharmonic, so its Hessian is nonnegative:

$$0 \le (u_{z_k \bar{z}_l}) = |e^{a(z)}|^2 (r_{z_k \bar{z}_l} + r_{z_k} \bar{a}_l + a_k r_{\bar{z}_l} + r a_k \bar{a}_l). \tag{23}$$

Recall that we are assuming r does not vanish. In (23) substitute $a_k = \frac{-r_{z_k}}{r}(z)$. After cancelling the positive factor $|e^{a(z)}|^2$ and multiplying by the positive number $r(z)$, we discover that

$$0 \le (r r_{z_k \bar{z}_l} - r_{z_k} r_{\bar{z}_l}) = M(r).$$

By Definition VI.4.5, this inequality yields the logarithmic plurisubharmonicity of r. $\square$

The ideas in Proposition VI.4.9 easily extend to convex increasing functions other than the exponential. Let ϕ be a smooth function on the real line, and suppose that both ϕ' and ϕ'' are positive. We next characterize those functions r for which $\phi^{-1}(r)$ is plurisubharmonic. In Proposition VI.4.10, we will write $g = \phi^{-1}(r)$. To simplify the notation we introduce an equivalence class: we say that $h \in E(g)$ if $H(h) = H(g)$.

Proposition VI.4.10. Let ϕ be a smooth convex increasing function on **R**. Suppose that g is a smooth function defined on some open set in $\mathbf{C}^n$. Then g is plurisubharmonic if and only if $\phi(h)$ is plurisubharmonic for every h in $E(g)$.

Proof. Suppose that g is plurisubharmonic and choose $h \in E(g)$. Since g and h have the same Hessian, h is also plurisubharmonic; by Exercise 13, $\phi(h)$ is plurisubharmonic.

To show the converse assertion we first suppose that $h \in E(g)$. We know that the Hessian $H(\phi(h))$ is nonnegative definite at a given point p:

$$0 \leq H(\phi \circ h) = \phi''(h)|\,\partial h\,|^2 + \phi'(h)H(h). \tag{24}$$

We now select $h \in E(g)$ such that ∂h vanishes at p. This choice is always possible; we choose h so that $g - h$ is twice the real part of a linear function, thereby simplifying (24). At p we see from (24) that the Hessian of the composition is $\phi'(h)H(h)$, which is a positive multiple of $H(h)$ at p. Thus $H(h)$ is nonnegative definite at p. Since g and h have the same Hessian, $H(g)$ is also nonnegative there. Since p is an arbitrary point, g is plurisubharmonic. $\qquad\square$

We now return to the special case where ϕ is the exponential function. The nonnegativity of $M(r)$ is equivalent to the nonnegativity of a *bordered* Hessian.

Definition VI.4.11. Let r be a smooth function of n complex variables. Its *bordered complex Hessian* $\mathcal{B}(r)$ is the $(n+1)$-by-$(n+1)$ Hermitian matrix

$$\begin{pmatrix} r_{z_1\bar{z}_1} & r_{z_1\bar{z}_2} & \cdots & r_{z_1\bar{z}_n} & r_{z_1} \\ r_{z_2\bar{z}_1} & r_{z_2\bar{z}_2} & \cdots & r_{z_2\bar{z}_n} & r_{z_2} \\ \cdots & \cdots & \cdots & \cdots & \cdots \\ r_{z_n\bar{z}_1} & \cdots & \cdots & r_{z_n\bar{z}_n} & r_{z_n} \\ r_{\bar{z}_1} & \cdots & \cdots & r_{\bar{z}_n} & r \end{pmatrix} = \begin{pmatrix} H(r) & \partial r \\ \overline{\partial} r & r \end{pmatrix}.$$

Lemma VI.4.12. A smooth nonnegative function is logarithmically plurisubharmonic if and only if its bordered Hessian is nonnegative definite.

Proof. In case r vanishes somewhere, its gradient also vanishes there, and $H(r)$ is nonnegative definite. At such points we have

$$\mathcal{B}(r) = \begin{pmatrix} H(r) & 0 \\ 0 & 0 \end{pmatrix}$$

and the bordered Hessian is evidently also nonnegative definite there. Also, $M(r)$ is zero, and hence also nonnegative definite. Thus when r vanishes both $M(r)$ and $\mathcal{B}(r)$ are nonnegative definite; thus there is nothing to prove at these points. We may therefore assume that r is positive.

Suppose first that $\mathcal{B}(r) \geq 0$. For all $a \in \mathbf{C}^n$, and for all $b \in \mathbf{C}$ we have

$$H(r)(a, a) + 2 \operatorname{Re}(\overline{b}\langle \partial r, \overline{a}\rangle) + r|b|^2 \geq 0.$$

Since r is positive, we may choose $b = -\frac{\langle \partial r, \overline{a}\rangle}{r}$. This choice yields

$$H(r)(a, a) - \frac{|\langle \partial r, \overline{a}\rangle|^2}{r} \geq 0.$$

It follows immediately that $M(r) \geq 0$.

Conversely, suppose that $M(r) \geq 0$. The elementary inequality

$$r 2 \operatorname{Re}(\overline{b}\langle \partial r, \overline{a}\rangle) \geq -|rb|^2 - |\langle \partial r, \overline{a}\rangle|^2$$

holds. Therefore, for all $a \in \mathbf{C}^n$, and for all $b \in \mathbf{C}$, we obtain

$$rH(r)(a, a) + 2r \operatorname{Re}(\overline{b}\langle \partial r, \overline{a}\rangle) + r^2|b|^2 \geq rH(r)(a, a) - |\langle \partial r, \overline{a}\rangle|^2$$

$$\geq 0,$$

and hence $r\mathcal{B}(r) \geq 0$. Since r is positive we see that $\mathcal{B}(r) \geq 0$ as well. $\square$

Lemma VI.4.13. Suppose that r is a twice differentiable bihomogeneous function. Then $\det(\mathcal{B}(r))$ vanishes identically.

Proof. Suppose r is bihomogeneous of degree $2m$. By differentiating (14) with respect to $\overline{\lambda}$ and then setting $\lambda = 1$, we obtain

$$\sum r_{\overline{z}_k}(z, \overline{z})\overline{z}_k = mr(z, \overline{z}). \tag{25}$$

Differentiating (25) with respect to z_j we further obtain, for $j = 1, \ldots, n$,

$$\sum r_{z_j \overline{z}_k}(z, \overline{z})\overline{z}_k = mr_{z_j}(z, \overline{z}). \tag{26}$$

It follows immediately from (25) and (26) that the column vector $(\overline{z}_1, \ldots, \overline{z}_n, -m)$ lies in the kernel of the matrix $\mathcal{B}(r)$. $\square$

Formula (25) is the analogue in several complex variables of a famous elementary result of Euler for homogeneous functions in several real variables. Differentiating Euler's formula gives a version of (26) in several variables as well. See Exercise 14.

We are now prepared to prove a surprising result. The conclusion holds for twice differentiable bihomogeneous functions of an arbitrary positive degree $2m$. We restrict our consideration here to polynomials.

Theorem VI.4.14. Suppose that r is a nonnegative bihomogeneous polynomial. Then r is logarithmically plurisubharmonic if and only if r is plurisubharmonic.

Proof. As we have noted in the discussion after Definition VI.4.5, if $\log(r)$ is plurisubharmonic, then so is r. We are interested here in the converse assertion; we assume that r is plurisubharmonic and we will prove that $\log(r)$ also is.

First observe that, for each $m \geq 1$, the bihomogeneous polynomial ϕ of degree $2m$ defined by $\phi(z, \overline{z}) = \|z\|^{2m}$ is strongly plurisubharmonic away from the origin. Given r, and $\epsilon > 0$, consider the function $r + \epsilon\phi$ which is plurisubharmonic away from the origin. We claim that $\mathcal{B}(r + \epsilon\phi) \geq 0$. It suffices to show two things: 1) the first n leading principal minor determinants are positive, and 2) the $(n+1)$-st leading principal minor determinant is nonnegative. The positivity of the first n minor determinants is equivalent to the strong plurisubharmonicity of $r + \epsilon\phi$, so the first statement holds. Since $r + \epsilon\phi$ is bihomogeneous, Lemma VI.4.13 guarantees that the determinant of its bordered Hessian vanishes identically. This proves the second statement, and hence $\mathcal{B}(r + \epsilon\phi)$ is nonnegative definite. Letting ϵ tend to zero we see that $\mathcal{B}(r)$ is also nonnegative definite. By Lemma VI.4.12, r is logarithmically plurisubharmonic, which is the desired conclusion. $\square$

Exercise 12. Suppose that r is a smooth function of one complex variable. Verify that subharmonicity is equivalent to the mean-value inequality discussed in Remark VI.4.3.

Exercise 13. Let f be a smooth function on a subset of $\mathbf{R}$, and let r be a smooth function on $\mathbf{C}^n$. Compute the Hessian of the composition $f \circ r$. Use this to show that, when r is smooth and nonnegative, $M(r) = r^2 H(\log(r))$. Let r be plurisubharmonic, and let f be convex and increasing. Show that $f \circ r$ is plurisubharmonic.

Exercise 14. Let r be a twice differentiable homogeneous function of several real variables. Find formulas analogous to (25) and (26).

Exercise 15. Fill in the details in the proofs of Example VI.4.8.

VI.5   Positivity conditions for polynomials

In this section we introduce eight positivity conditions for real-valued polynomials on $\mathbf{C}^n$. Each of the eight conditions arises naturally in our discussion, and seven of them turn out to be distinct. We discuss the various implications involving these conditions. We provide easy proofs of many of the implications, and we will provide examples revealing that the converse implications fail in general. Most of the converse assertions fail even for bihomogeneous polynomials.

Definition VI.5.1. (Positivity Conditions) Let $r : \mathbf{C}^n \times \mathbf{C}^n \to \mathbf{C}$ be a polynomial such that r is not identically zero and $r(z, \overline{z})$ is real for all z. We introduce the following eight conditions:

P1) On the diagonal, r is nonnegative as a function. That is, for all z,

$$r(z, \overline{z}) \geq 0.$$

P2) On the diagonal, r is the quotient of squared norms of holomorphic polynomial mappings. That is, there is an N and holomorphic polynomial mappings f and g from $\mathbf{C}^n$ to $\mathbf{C}^N$ such that

$$r(z, \overline{z}) = \frac{\| f(z) \|^2}{\| g(z) \|^2}.$$

P3) On the diagonal, r is the squared norm of a holomorphic polynomial mapping. That is, there is an N and a holomorphic polynomial mapping f from $\mathbf{C}^n$ to $\mathbf{C}^N$ such that

$$r(z, \overline{z}) = \| f(z) \|^2.$$

P4) The matrix of coefficients of r is nonnegative definite. By Theorem IV.5.2 this is the same as saying the matrix of coefficients of r is of the form A^*A. That is,

$$r(z, \overline{z}) = \sum c_{\alpha\beta} z^\alpha \overline{z}^\beta$$

and the entries $c_{\alpha\beta}$ of the (necessarily) Hermitian matrix C satisfy

$$c_{\alpha\beta} = \langle f_\alpha, f_\beta \rangle$$

for certain vectors f_α.

P5) r is positive at one point on the diagonal, and there is a positive integer m such that, on the diagonal, r^m is a squared norm of a holomorphic polynomial mapping. That is, there is an N and a holomorphic polynomial mapping $f : \mathbf{C}^n \to \mathbf{C}^N$ such that

$$r(z, \overline{z})^m = \| f(z) \|^2.$$

P6) r is positive at one point on the diagonal, and r satisfies the global Cauchy-Schwarz inequality: for all z and w,

$$r(z, \overline{z})r(w, \overline{w}) \geq |r(z, \overline{w})|^2. \tag{27}$$

P7) On the diagonal, r is nonnegative and logarithmically plurisubharmonic. That is, $M(r) \geq 0$; for each $a \in \mathbf{C}^n$ we have

$$rH(r)(a, a) \geq |\langle \partial r, \overline{a} \rangle|^2$$

at every point.

P8) On the diagonal, r is plurisubharmonic. That is, for all $a \in \mathbf{C}^n$,

$$H(r)(a, a) \geq 0$$

at every point.

Remark VI.5.2. We could have stated these conditions in greater generality, but more general situations are not needed in the rest of the book. All eight conditions make sense for real-analytic real-valued functions, and most of the conditions make sense for smooth real-valued functions. Some make sense for continuous functions.

Remark VI.5.3. P3) and P4) are equivalent. One implication follows for example from Theorem IV.5.2. The other is easy. If $c_{\alpha\beta} = \langle f_\alpha, f_\beta \rangle$, then

$$\sum_{\alpha,\beta} c_{\alpha\beta} z^\alpha \overline{z}^\beta = \sum_{\alpha,\beta} \langle f_\alpha, f_\beta \rangle z^\alpha \overline{z}^\beta = \left\| \sum_\alpha f_\alpha z^\alpha \right\|^2.$$

We next discuss the implications among the positivity conditions. It is immediate that P3) implies P2) implies P1), and that P3) implies P5). We observed above that P3) and P4) are equivalent. Our subsequent examples show that no other converse assertion holds!

Lemma VI.5.4. P5) implies P6) implies P7) implies P8).

Proof. Suppose P5) holds, and write $r^m = \| f \|^2$. When m is odd, we see that r is nonnegative. When m is even, our hypotheses that r is positive at a single point and that r is a polynomial, in combination with $r^m = \| f \|^2$, guarantee also that r is nonnegative. Then, by the usual Cauchy-Schwarz inequality,

$$(r(z, \overline{z}) r(w, \overline{w}))^m = \| f(z) \|^2 \| f(w) \|^2$$
$$\geq | \langle f(z), f(w) \rangle |^2 = | r(z, \overline{w}) |^{2m}. \quad (28)$$

Since $m > 0$, we may take m-th roots of both sides of (28) and preserve the direction of the inequality. Thus (27) and hence P6) hold.

Next suppose P6) holds. Choose w so that $r(w, \overline{w}) > 0$; then (27) implies $r(z, \overline{z}) \geq 0$ for all z. To verify P7) it remains to show that $M(r) \geq 0$. Define a function u by

$$u(z, \overline{z}) = r(z, \overline{z}) r(w, \overline{w}) - | r(z, \overline{w}) |^2.$$

According to P6), the function u is nonnegative, and it evidently vanishes at w. Therefore $H(u)$ must be nonnegative definite at w. Thus:

$$0 \leq \sum u_{z_j \overline{z}_k}(w, \overline{w}) a_j \overline{a}_k = H(u)(a, a).$$

The rules of elementary calculus express the derivatives of u in terms of the derivatives of r; evaluating at $z = w$ yields $H(u) = M(r)$. Therefore $M(r) \geq 0$ and hence P7) holds. We note in passing that we could obtain the same result by taking logarithms of both sides of the inequality before differentiating.

That P7) implies P8) was noted in the discussion after Definition VI.4.5. We repeat the argument. Assuming that $M(r) \geq 0$ we want to show that $H(r) \geq 0$. Where r is positive, this conclusion follows from the formula $r H(r) = M(r) + | \partial r |^2$. Where r vanishes, it is at a minimum point. Therefore $H(r) \geq 0$ there by the second derivative test. $\square$

Gathering this information together shows that P3) implies all the other properties. Thus being a squared norm of a holomorphic polynomial mapping is a strong property for a real-valued polynomial. On the other hand, the conditions other than P4) can hold for r even when it is not a squared norm.

The reader may be curious why P5) and P6) include an assumption on the value of r at one point. The reason is simple; without this assumption we could replace r by $-r$ and the important parts of P5) and P6) would be unchanged. For example, (27) implies that the values of r must be of one sign; we want the values to be nonnegative.

Our next example reveals that most of these conditions are distinct, even for bihomogeneous polynomials.

Example VI.5.5. For a real number a, define $r_a : \mathbf{C}^n \to \mathbf{R}$ by

$$r_a(z, \overline{z}) = \| z \|^{4n} - a \prod_{j=1}^{n} | z_j |^4.$$

We consider values of a for which the eight positivity conditions hold. For simplicity we assume that $n = 2$. The values obtained in the next result will change if we change the dimension.

Proposition VI.5.6. For $r_a(z, \overline{z}) = (| z_1 |^2 + | z_2 |^2)^4 - a| z_1 z_2 |^4$, the following eight statements are true:

P1) holds if and only if $a \leq 16$.

P2) holds if and only if $a < 16$.

P3) holds if and only if $a \leq 6$.

P4) holds if and only if $a \leq 6$.

P5) holds for $a \leq 7.8$, and fails for $a > 8$.

P6) holds if and only if $a \leq 8$.

P7) holds if and only if $a \leq 12$.

P8) holds if and only if $a \leq 12$.

Proof. Several of the calculations are facilitated by writing $x = | z_1 |^2$ and $y = | z_2 |^2$, and replacing r_a by $(x + y)^4 - ax^2y^2$. When $a < 0$, r_a

is a squared norm, so P3) holds, and all the statements are evident. So we assume that $a \geq 0$. We have

$$(x + y)^4 - ax^2 y^2 = ((x + y)^2 + \sqrt{a}\,xy)((x + y)^2 - \sqrt{a}\,xy). \quad (29)$$

Formula (29) shows that r_a is nonnegative if and only if the second factor in (29) is nonnegative. The condition is $2 - \sqrt{a} \geq -2$. Hence r_a is nonnegative if and only if $a \leq 16$, and hence the first statement is true.

The same calculation shows that r_a is positive away from the origin when $a < 16$.

One can verify that P2) holds for $a < 16$ by using Proposition VI.1.3. See Exercise 17. This also follows from Theorem VII.1.1; since r_a is bihomogeneous and positive away from the origin, there is an integer d and a holomorphic polynomial mapping h such that $r_a(z, \overline{z}) = (\|\,h(z)\,\|^2)/\|z\|^{2d}$. Since $\|z\|^{2d} = \|H_d(z)\|^2$, P2) holds if $a < 16$.

To finish the determination of when P2) holds, it remains to show that r_a is *not* a quotient of squared norms when $a = 16$. To do so, we show that the jet pullback property fails when $a = 16$. Let $z(t) = (1 + t, 1)$. Then

$$z^* r(t, \overline{t}) = (\,|\,1 + t\,|^2 + 1)^4 - 16|\,1 + t\,|^4$$
$$= 8(t^2 + 2|\,t\,|^2 + \overline{t}^2) + \cdots,$$

and the lowest order part does not satisfy the necessary condition. Thus r_a is not a quotient of squared norms when $a = 16$. One can also note that the zero-set of r_{16} is not a complex analytic variety.

It is easy to determine when P3) and P4) hold. They are equivalent by Remark VI.5.3, so we consider P3). Expanding $(x + y)^4$ by the binomial theorem shows that the coefficient of $x^2 y^2$ in r_a is $6 - a$ and all other coefficients are positive integers. Thus the condition for r_a being a squared norm is that $a \leq 6$.

Next we discuss P5). A power of r_a will be a squared norm if and only if all the coefficients of that power of $(x + y)^4 - ax^2 y^2$ are nonnegative. Computing r_a^2 gives

$$r_a^2 = x^8 + 8x^7y + (28 - 2a)x^6y^2 + (56 - 8a)x^5y^3$$
$$+ (70 - 12a + a^2)x^4y^4 + \cdots$$

This expression is symmetric in x and y, so to make r_a^2 a squared norm we need only make the coefficients of x^6y^2, of x^5y^3, and of x^4y^4 nonnegative. Hence r_a^2 is a squared norm when $a \leq 7$.

Using Mathematica with $a = 7.8$, one computes r_a^{32} and sees that all the coefficients are positive. Thus $r_{7.8}^{32}$ is a squared norm. By using Mathematica, one can see that r_a^{32} is not a squared norm when $a = 7.9$.

Comment. The author believes that P5) holds if and only if $a < 8$. It is fairly easy to prove (in this example) that the set of a for which P5) holds is an open set, and that the set of a for which P6) holds is a closed set. In particular P5) and P6) are distinct conditions. It is perhaps possible to prove that P5) and a sharp form of P6) are equivalent.

Next we show that r_a fails to satisfy P6) for $a > 8$. We evaluate both sides of the Cauchy-Schwarz inequality (27) at the points

$$z = \left(\frac{1}{\sqrt{2}}, \frac{1}{\sqrt{2}}\right) \quad \text{and} \quad w = \left(\frac{1}{\sqrt{2}}, \frac{-1}{\sqrt{2}}\right).$$

Then $r_a(z, \overline{w}) = \frac{-a}{16}$ and $r_a(z, \overline{z}) = r_a(w, \overline{w}) = 1 - \frac{a}{16}$. Plugging this result into the inequality (27) yields

$$\left|\frac{-a}{16}\right|^2 \leq \left(1 - \frac{a}{16}\right)^2.$$

This condition implies $1 - \frac{a}{8} \geq 0$, so the global Cauchy-Schwarz inequality implies $a \leq 8$. We omit the proof of the inequality when $a \leq 8$.

To prove P8), we use the special form of r_a to find its complex Hessian. We then use Theorem IV.5.7 (and lots of computation) to decide when r_a is plurisubharmonic. Let λ_{jk} denote the partial derivative $\frac{\partial}{\partial z_j} \frac{\partial}{\partial \overline{z}_k}(r_a)$. The entries of the Hessian follow:

$$\lambda_{11} = 4(|z_1|^2 + |z_2|^2)^3 + 12(|z_1|^2 + |z_2|^2)^2|z_1|^2$$
$$- 4a|z_1|^2|z_2|^4$$
$$\lambda_{12} = \overline{z}_1 z_2 (12(|z_1|^2 + |z_2|^2)^2 - 4a|z_1|^2|z_2|^2)$$
$$\lambda_{22} = 4(|z_1|^2 + |z_2|^2)^3) + 12(|z_1|^2 + |z_2|^2)^2|z_2|^2$$
$$- 4a|z_1|^4|z_2|^2.$$

By the bihomogeneity it suffices to determine the positive definiteness of the Hessian matrix on the unit sphere. The entries of the Hessian simplify:

$$\lambda_{11} = 4 + 12|z_1|^2 - 4a|z_1|^2|z_2|^4$$
$$\lambda_{12} = \overline{z}_1 z_2 (12 - 4a|z_1|^2|z_2|^2)$$
$$\lambda_{22} = 4 + 12|z_2|^2 - 4a|z_1|^4|z_2|^2.$$

Computing the determinant $\det \lambda = \lambda_{11}\lambda_{22} - |\lambda_{12}|^2$ and eliminating $|z_2|^2$ we find that

$$\det \lambda = 64 - 64a|z_1|^2 + 256a|z_1|^4 - 384a|z_1|^6 + 192a|z_1|^8. \quad (30)$$

We know that r_a is strongly plurisubharmonic if and only if both λ_{11} and the determinant $\det \lambda$ are positive. When $a = 12$ one can verify that the determinant equals

$$64(1 - 6|z_1|^2 + 6|z_1|^4)^2,$$

and this expression vanishes at two values in the unit interval. Again when $a = 12$ we see that $\lambda_{11} > 0$ for $0 \leq |z_1|^2 \leq 1$. The expression for λ_{11} is a cubic in $|z_1|^2$ with only one real zero, which lies outside the unit interval. Hence r_a is plurisubharmonic when $a = 12$; it follows easily that the same is true when $a \leq 12$. When $a < 12$, expression (30) is positive for $0 \leq |z_1| \leq 1$. When $a > 12$, this expression is negative for points close to its roots. Thus $a = 12$ is the cut-off point for plurisubharmonicity, and P8) holds.

The condition for P7) is the same as the condition for P8), according to Theorem VI.4.14. In Exercise 16 we ask the reader to verify this directly.   □

Remark VI.5.7. We have already given polynomials (Examples VI.4.7 and VI.4.8) for which conditions P7) and P8) differ. In Example VI.4.7, conditions P3), P5), P6) and P7) hold for the same values of the parameter c, namely only when $c = 0$, whereas P8) holds for all values of c.

Exercise 16. Verify directly that P7) holds in Proposition VI.5.6 if and only if $a \leq 12$.

Exercise 17. Verify that P2) holds in Proposition VI.5.6 when $a < 16$ by using Proposition VI.1.3. Suggestion: The polynomial q given by $q(t) = (t + 1)^4 - at^2$ satisfies the hypotheses of Proposition VI.1.3, and therefore its conclusion. Homogenize the result.

Exercise 18. (Difficult) Prove that P6) holds in Proposition VI.5.6 for $a = 8$. It follows that P6) holds if and only if $a \leq 8$.

(Open problem) Prove that P5) holds in Proposition VI.5.6 if and only if $a < 8$.

CHAPTER VII

Stabilization and Applications

The results in this chapter generally have the same flavor. We are given a polynomial whose values are positive on some set. We show that the polynomial agrees with a quotient of squared norms on that set, thus explaining the positivity. These results are striking applications of Theorem VII.1.1; this stabilization result shows that certain bihomogeneous polynomials are quotients of squared norms. See Remark VII.1.4 for an explanation of the term *stabilization*.

We state and discuss Theorem VII.1.1 in Section 1, but postpone its proof to Section 6. The proof uses the Hilbert space methods developed in this book; it combines properties of the Bergman kernel function for the unit ball in $\mathbf{C}^n$ with results about compact operators.

VII.1 Stabilization for positive bihomogeneous polynomials

Recall that the term *squared norm* means for us a *finite* sum of squared absolute values of holomorphic polynomial functions. We have observed that a power of the squared Euclidean norm is itself a squared norm; $|| z ||^{2d} = || H_d(z) ||^2 = || z^{\otimes d} ||^2$. In this section we state and discuss Theorem VII.1.1. This result implies that a bihomogeneous polynomial, whose values away from the origin are positive, is necessarily a quotient of squared norms. We may always choose the denominator to be $|| H_d(z) ||^2$ for some d.

Let r be a real-valued bihomogeneous polynomial of degree $2m$ with matrix of coefficients $(c_{\alpha\beta})$. Recall that V_m denotes the vector space of (holomorphic) homogeneous polynomials of degree m. According to Proposition VI.2.10, r determines a Hermitian form on V_m defined by

$$\langle Lh, g \rangle = \sum_{\alpha,\beta} c_{\alpha\beta} \langle h, z^\beta \rangle \overline{\langle g, z^\alpha \rangle}. \tag{1}$$

Furthermore, this form is positive definite if and only if r is itself a squared norm of a polynomial mapping whose components form a basis of V_m. More generally the form is nonnegative if and only if r is a squared norm.

Suppose that r is a squared norm, say $r = \| f \|^2$. Then r will be positive away from the origin if and only if the set $\mathbf{V}(f)$ of common zeroes of the components of f consists of the origin alone. The more interesting situation arises when r is positive away from the origin, but r is not a squared norm.

The following decisive theorem clarifies the issue. A bihomogeneous polynomial is positive away from the origin precisely when it is a quotient of squared norms of holomorphic polynomial mappings, and both the numerator and denominator vanish only at the origin. This statement is S6) below.

Theorem VII.1.1. **(Catlin-D'Angelo, Quillen) (Stabilization in the bihomogeneous case)** Let r be a real-valued bihomogeneous polynomial of degree $2m$. The following six statements are equivalent.

S1) $r(z, \overline{z}) > 0$ for $z \neq 0$.

S2) The minimum value of r on the unit sphere is a positive number.

S3) There is an integer d such that the Hermitian matrix for $\| z \|^{2d}$ $r(z, \overline{z})$ is positive definite. Thus

$$\| z \|^{2d} r(z, \overline{z}) = \sum E_{\mu\nu} z^\mu \overline{z}^\nu \tag{2}$$

where $(E_{\mu\nu})$ is positive definite.

S4) Let $k_d(z, \zeta) = \langle z, \zeta \rangle^d r(z, \overline{\zeta})$. There is an integer d such that the integral operator defined on $A^2(\mathbf{B}_n)$ by the kernel k_d is positive definite from V_{m+d} to itself.

S5) There is an integer d and a holomorphic homogeneous vector-valued polynomial g of degree $m + d$ such that $\mathbf{V}(g) = \{0\}$ and such that $\| z \|^{2d} r(z, \overline{z}) = \| g(z) \|^2$.

S6) r is a quotient $(\| g \|^2)/(\| h \|^2)$ of squared norms of holomorphic homogeneous polynomial mappings whose zero-sets $\mathbf{V}(g)$ and $\mathbf{V}(h)$ consist of the origin alone.

It is valuable to compare Theorem VII.1.1 with our results from Hermitian linear algebra. Corollary IV.5.1 and Theorem IV.5.7 give methods for deciding whether Hermitian forms (that is, bihomogeneous polynomials of degree 2) are positive definite (that is, have positive values away from the origin). Theorem VII.1.1 gives a method for showing that a bihomogeneous polynomial r of arbitrary even degree has positive values away from the origin. One can apply these results from Chapter IV to the matrix of coefficients of the polynomial $\| z \|^{2d} r(z, \overline{z})$, for d sufficiently large. Warning VII.1.5 below reveals that one has no control on the size of d in general. If r has degree 2, then we may take $d = 0$. When r has degree 4 or more, however, there is no bound on the minimum d that works. Theorem VII.1.1 nevertheless provides a useful test of positivity.

We have already established many of the implications in Theorem VII.1.1. The equivalence of S1) and S2) follows immediately from the homogeneity of r. We verified the equivalence of S3) and S4) in Proposition VI.2.10.

Suppose $(E_{\mu\nu})$ from (2) is positive definite. Corollary VI.2.7 implies that the right-hand side of (2) is a squared norm of a holomorphic polynomial mapping whose components form a basis for V_{m+d}. In this case the right-hand side of (2) vanishes only at the origin, and hence S3) implies S5).

Recall that $\| z \|^{2d} = \| H_d(z) \|^2$; this function vanishes only at the origin. Therefore S5) implies S6), which tells us that r is a quotient of squared norms. Since $\mathbf{V}(g)$ and $\mathbf{V}(h)$ consist of the origin alone, r is positive away from the origin. Hence S6) implies S1).

Taking these known implications into account, to prove Theorem VII.1.1 it suffices to show that S1) implies S4). We will accomplish this by using some of the facts about compact operators and the Bergman kernel we have already proved.

We prove the remaining implication, S1) implies S4), in Section 6. Before proceeding to the applications we close this section by making a few more remarks and deriving Pólya's Theorem from Theorem VII.1.1.

Remark VII.1.2. Recall the positivity conditions from Definition VI.5.1. For bihomogeneous polynomials, Theorem VII.1.1 shows that a sharp form of P1) implies P2). Even for bihomogeneous polynomials, P1) does not imply P2), as Proposition VI.5.6 shows when $a = 16$.

Remark VII.1.3. The minimum d required in S5) may be smaller than the minimum d required in S3). For example, suppose $r(z, \overline{z}) = |z_1|^4 + |z_2|^4$. Then S5) is true for $d = 0$, while S3) requires $d \geq 1$.

Remark VII.1.4. For each positive integer k, we have

$$\| z \|^{2k} \| g \|^2 = \| H_k \otimes g \|^2.$$

From this we have the following *stabilization* property. If S5) holds for some d_0, then S5) holds for each d with $d \geq d_0$. Similarly, if S3) or S4) holds for some d_0, then it holds for each d with $d \geq d_0$.

Warning VII.1.5. The smallest d required in S5) can be arbitrarily large even for polynomials of degree 4 in two variables. Example VI.3.1, where r_c is given by $|z_1|^4 + c|z_1 z_2|^2 + |z_2|^4$, shows this. The condition for the positivity of r_c away from the origin is that $c > -2$. In Exercise 1 the reader is asked to show that the minimum d_c that works in S5) tends to infinity as c tends to -2. In the general case, one can prove estimates for the minimum d using other information about r, such as its minimum and maximum values on the unit sphere.

Observe that the polynomial in this warning depends upon the absolute values of the complex variables, but not on their arguments. By restricting to general bihomogeneous polynomials with this property we obtain Pólya's result (Theorem VI.1.4) as a corollary.

Corollary VII.1.6. (Pólya) Let p be a homogeneous polynomial in n real variables of degree m with real coefficients. Suppose that $p(x) > 0$ for all nonzero x in the set determined by $x_j \geq 0$ for all j. Then there is an integer d and a homogeneous polynomial q of degree $m + d$ with positive coefficients such that

$$s(x)^d p(x) = (x_1 + \cdots + x_n)^d p(x) = q(x).$$

Proof. Let p be a homogeneous polynomial on $\mathbf{R}^n$ satisfying the hypotheses of Pólya's Theorem. For $j = 1, \ldots, n$ we substitute $x_j = |z_j|^2$ and obtain a corresponding bihomogeneous polynomial on $\mathbf{C}^n$, written $p(z, \overline{z})$; evidently $p(z, \overline{z}) > 0$ for $z \neq 0$. According to statement S3) from Theorem VII.1.1, there is an integer d and a positive definite matrix $(E_{\mu\nu})$ such that

$$s(x)^d p(x) = \| z \|^{2d} p(z, \overline{z}) = \sum E_{\mu\nu} z^\mu \overline{z}^\nu.$$

The middle term depends only upon the $|z_j|^2$; hence so does the right-hand side. It follows that the $E_{\mu\nu}$ there vanish unless $\mu = \nu$. We therefore obtain

$$s(x)^d p(x) = \sum E_{\mu\mu} |z^\mu|^2 = \sum E_{\mu\mu} x^\mu,$$

where the coefficients $E_{\mu\mu}$ are the eigenvalues of a positive definite matrix and hence are positive numbers. $\qquad\square$

Exercise 1. Let r_c be as in Warning VII.1.5. Explicitly compute $(|z_1|^2 + |z_2|^2)^d r_c(z)$ and determine the condition on d and c that makes this product a squared norm. Then verify that the minimum d tends to infinity as c tends to -2. Suggestion: Write $x = |z_1|^2$ and $y = |z_2|^2$ and work only with the real variables.

The main assertion that S1) and S3) from Theorem VII.1.1 are equivalent was proved in 1967 by Quillen. Unaware of that result, Catlin and the author, motivated by trying to prove Theorem VII.3.1, found a different proof, which we give in Section VII.6. Both proofs use hard analysis; Quillen uses Gaussian integrals on all of $\mathbf{C}^n$, whereas the proof here uses the Bergman kernel function on the unit ball $\mathbf{B}_n$. Quillen uses a priori estimates, whereas the proof here uses facts about compact operators on $L^2(\mathbf{B}_n)$.

VII.2 Positivity everywhere

In this section we consider positivity conditions for polynomials that are not bihomogeneous. Let $r : \mathbf{C}^n \times \mathbf{C}^n \to \mathbf{C}$ be an arbitrary polynomial. As usual we consider it on the diagonal. Suppose that r is bounded away from 0 on the diagonal; thus there is $\epsilon > 0$ such that

$$r(z, \overline{z}) \geq \epsilon > 0$$

for all z. Must r be a quotient of squared norms? The answer is no. The following example, which appeared earlier as Example VI.3.5, recalls for us the necessity of the jet pullback property.

Example VII.2.1. Let $n = 1$ and put

$$r(z, \overline{z}) = 1 + z^2 + \overline{z}^2 + 2|z|^2.$$

Since $|\operatorname{Re}(z^2)| \leq |z|^2$, we see that $r(z, \overline{z}) \geq 1$ for all z. On the other hand, we showed in Example VI.3.5 that the jet pullback property fails, and hence r is not the quotient of squared norms.

The author does not know whether the jet pullback property plus an appropriate nonnegativity condition together provide a sufficient condition for r to be the quotient of squared norms. In our next theorem we will place an additional restriction on r and conclude that r is a quotient of squared norms. First we introduce some notation to make the proof more transparent. Suppose that r is a real-valued polynomial in z and $\overline{z}$. Then its degree in z, say m, equals its degree in $\overline{z}$. Then r is of degree at most $2m$, but it is not necessarily of degree $2m$. In Example VII.2.1, for instance, r is of degree 2 whereas $2m = 4$. We let r_h denote the polynomial whose terms are those terms in r of degree exactly $2m$. In Example VII.2.1, r_h vanishes identically.

The next result shows that positivity for r and r_h together yield an expression for r as a quotient of squared norms.

Theorem VII.2.2. Let r be a polynomial such that $r(z, \overline{z}) \geq \epsilon > 0$ for all $z \in \mathbf{C}^n$. Suppose in addition that the bihomogeneous polynomial r_h is positive away from the origin. Then r is a quotient of squared norms of holomorphic polynomial mappings. Thus there exist an integer N and holomorphic polynomial mappings $A : \mathbf{C}^n \to \mathbf{C}^N$ and $B : \mathbf{C}^n \to \mathbf{C}^N$ such that A and B vanish nowhere and

$$r(z, \overline{z}) = \frac{\| A(z) \|^2}{\| B(z) \|^2}.$$

Proof. Given r, we define a polynomial R in the $n + 1$ variables z and t by

$$R(z, t, \overline{z}, \overline{t}) = | t |^{2m} r \left(\frac{z}{t}, \frac{\overline{z}}{\overline{t}} \right) \tag{3}$$

when $t \neq 0$ and by

$$R(z, 0, \overline{z}, 0) = r_h(z, \overline{z}). \tag{4}$$

The purpose of (3) and (4) is to bihomogenize r. Note that $R(z, 1, \overline{z}, 1) = r(z, \overline{z})$. Since

$$R(z, t, \overline{z}, \overline{t}) = \sum_{0 \leq | \alpha |, | \beta | \leq m} c_{\alpha\beta} z^\alpha t^{m - | \alpha |} \overline{t}^{m - | \beta |} \overline{z}^\beta,$$

we see that R is a bihomogeneous polynomial in its variables. We claim that it assumes only positive values away from the origin in $\mathbf{C}^{n+1}$. For $t = 0$, this is true by (4), since r_h is assumed positive on the unit sphere.

Suppose that $r_h \geq c > 0$ there. Hence there is a $\delta > 0$ such that the values of R are at least $\frac{c}{2}$ for $\| z \|^2 + | t |^2 = 1$ and $| t | \leq \delta$. For $| t | \geq \delta$ and $\| z \|^2 + | t |^2 = 1$, (3) guarantees that

$$R(z, t, \overline{z}, \overline{t}) = | t |^{2m} r \left(\frac{z}{t}, \frac{\overline{z}}{\overline{t}} \right) \geq \delta^{2m} \epsilon.$$

Thus $R > 0$ on the unit sphere.

By Theorem VII.1.1 there is an integer d and a polynomial mapping h such that

$$(\| z \|^2 + | t |^2)^d R(z, t, \overline{z}, \overline{t}) = \| h(z, t) \|^2.$$

Setting $t = 1$ gives us

$$(\| z \|^2 + 1)^d r(z, \overline{z}) = \| h(z, 1) \|^2.$$

Since $(\| z \|^2 + 1)^d$ is a squared norm of a holomorphic mapping $B(z)$, we conclude that r is the quotient of squared norms $(\| A \|^2)/(\| B \|^2)$. Furthermore $\| B(z) \|^2 \geq 1$ for all z; since $r \geq \epsilon > 0$, we obtain

$$\| A(z) \|^2 = \| B(z) \|^2 r(z, \overline{z}) \geq \epsilon.$$

Therefore also $A(z)$ is never 0. $\qquad \square$

Example VII.2.3. A polynomial can be positive everywhere without being bounded away from zero. One simple example is given by

$$r(z, \overline{z}) = | z_1 z_2 - 1 |^2 + | z_1 |^2.$$

This r is never zero. Consider the curve given by $z(t) = (\frac{1}{t}, t)$. Then $z^* r(t) = | \frac{1}{t} |^2$ which tends to zero at infinity.

It is natural to ask what happens if we weaken the condition in Theorem VII.2.2 that r is bounded away from zero to r is positive. The proof breaks down, and the author does not know a general theorem in this case. Let us reconsider the proof without making the assumption that r is bounded away from zero. The first step is to homogenize.

Suppose we can then prove that

$$R(z, t, \overline{z}, \overline{t}) = \frac{\| A(z, t) \|^2}{\| B(z, t) \|^2}$$

is a quotient of squared norms. Then, since $r(z, \overline{z}) = R(z, 1, \overline{z}, 1)$, we see that

$$r(z, \overline{z}) = \frac{\| A(z, 1) \|^2}{\| B(z, 1) \|^2}.$$

Hence r is a quotient of squared norms. On the other hand, we see that

$$r_h(z, \overline{z}) = \frac{\| A(z, 0) \|^2}{\| B(z, 0) \|^2}.$$

Thus the homogenization technique in the proof proves that both r and r_h are quotients of squared norms. When the homogenization technique works, we obtain a holomorphic homotopy between r and r_h.

VII.3   Positivity on the unit sphere

A bihomogeneous polynomial is determined by its values on the unit sphere. In this section we consider (not necessarily bihomogeneous) real-valued polynomials on $\mathbf{C}^n$ that are positive on the unit sphere. We prove that such a polynomial agrees with the squared norm of a (holomorphic) polynomial mapping there. We then give some applications to proper mappings between balls.

Let r be a real-valued polynomial on $\mathbf{C}^n$, and suppose $r(z, \overline{z}) > 0$ for $\| z \|^2 = 1$. We cannot expect r to be a squared norm! Consider $r(z, \overline{z}) = 2 - \| z \|^2$.

Exercise 2. Prove that $2 - \| z \|^2$ cannot agree with a squared norm on any open set.

The proof of Theorem VII.3.1 below relies on Theorem VII.1.1. We homogenize r by adding a variable, as we did in Theorem VII.2.2.

We then add a correction term so that the resulting polynomial will be bihomogeneous and positive away from the origin. We can then apply Theorem VII.1.1.

Theorem VII.3.1. Suppose that $z \to r(z, \overline{z})$ is a real-valued polynomial on $\mathbf{C}^n$, and suppose that $r(z, \overline{z}) > 0$ for $\| z \| = 1$. Then there is an integer N and a holomorphic polynomial mapping $h : \mathbf{C}^n \to \mathbf{C}^N$ such that, for $\| z \| = 1$,

$$r(z, \overline{z}) = \| h(z) \|^2 = \sum_{j=1}^{N} | h_j(z) |^2.$$

Proof. We write

$$r(z, \overline{z}) = \sum_{|\alpha|,|\beta|=0}^{m} c_{\alpha\beta} z^\alpha \overline{z}^\beta.$$

Since r is real-valued, we have $c_{\alpha\beta} = \overline{c_{\beta\alpha}}$ by Proposition VI.2.1. By multiplying r by $\| z \|^2$ if necessary, we may assume that m is even without changing the hypotheses.

Let t be a complex variable. For a suitable positive constant C we define a bihomogeneous polynomial F_C on $\mathbf{C}^n \times \mathbf{C}$ by

$$F_C(z, t, \overline{z}, \overline{t}) = C(\| z \|^2 - | t |^2)^m + \sum c_{\alpha\beta} z^\alpha \overline{z}^\beta t^{m-|\alpha|} \overline{t}^{m-|\beta|} \quad (5)$$

We claim that the summation in (5) is positive when $\| z \|^2 = | t |^2 \neq 0$. To see this claim, observe (as in the proof of Theorem VII.2.2) that the summation equals $| t |^{2m} r(\frac{z}{t}, \frac{\overline{z}}{\overline{t}})$ and hence is positive when $\| \frac{z}{t} \|^2 = 1$. On the other hand, the summation is continuous in both z and t. It is positive when $\| z \|^2 = | t |^2 = \frac{1}{2}$, and hence there is a $\delta > 0$ and a positive number k such that

$$| t |^{2m} r \left(\frac{z}{t}, \frac{\overline{z}}{\overline{t}} \right) \geq k > 0$$

when $| (\| z \|^2 - | t |^2) | < \delta$ and $\| z \|^2 + | t |^2 = 1$. Therefore, on the unit sphere in $\mathbf{C}^{n+1}$ and for $| (\| z \|^2 - | t |^2) | < \delta$, we have

$$F_C(z, t, \overline{z}, \overline{t}) \geq k.$$

When $| (\| z \|^2 - | t |^2) | \geq \delta$ we have

$$F_C(z, t, \overline{z}, \overline{t}) \geq C\delta^m + \sum c_{\alpha\beta} z^\alpha \overline{z}^\beta t^{m-|\alpha|} \overline{t}^{m-|\beta|}. \tag{6}$$

The summation in (6) is a polynomial and therefore continuous; hence it achieves a minimum value η on the unit sphere in $\mathbf{C}^{n+1}$. Note that η can be negative. If we choose C so that $C\delta^m + \eta$ is positive, then the bihomogeneous polynomial F_C will be positive on the unit sphere. By Theorem VII.1.1, there is an integer d and a holomorphic polynomial mapping $h(z, t)$ so that

$$(\| z \|^2 + | t |^2)^d F_C(z, t, \overline{z}, \overline{t}) = \| h(z, t) \|^2$$

holds everywhere. Setting $t = 1$ and then $\| z \|^2 = 1$ shows that

$$2^d r(z, \overline{z}) = \| h(z, 1) \|^2$$

on the unit sphere in $\mathbf{C}^n$, which completes the proof.   □

VII.4 Applications to proper holomorphic mappings between balls

A continuous map $f : X \to Y$ between topological spaces is called *proper* if $f^{-1}(K)$ is compact in X whenever K is compact in Y. Our interest here will be proper holomorphic mappings between open unit balls in complex Euclidean spaces, perhaps of different dimensions. The model example of a proper holomorphic mapping is the function $z \mapsto z^m$ on $\mathbf{B}_1$; here m is a positive integer. The mapping H_m defined by (12) in Chapter VI is the analogous proper mapping in higher dimensions; H_m maps $\mathbf{B}_n$ properly to $\mathbf{B}_N$, where $N = d(m, n)$.

We begin with a simple characterization of proper continuous mappings between bounded domains. Recall that a domain is an open and connected set. We say that a sequence $\{z_\nu\}$ in Ω tends to the boundary of Ω if the distance from z_ν to the boundary of Ω tends to zero. The sequence $\{z_\nu\}$ itself need not converge. The reader who knows about the *compactification* of a space can reformulate the following result using that terminology.

Proposition VII.4.1. Let $\Omega \subset \mathbf{C}^n$ and $\Omega' \subset \mathbf{C}^N$ be bounded domains. A continuous mapping $f : \Omega \to \Omega'$ is proper if and only if the following condition holds. If $\{z_\nu\}$ is a sequence of points in Ω and tending to the boundary of Ω, then the image sequence $\{f(z_\nu)\}$ tends to the boundary of Ω'.

Proof. It is easy to prove the contrapositive of each required statement. If the condition fails, then there is some sequence $\{z_\nu\}$ tending to the boundary whose image does not. Hence there is a subsequence whose image stays within a compact set in the target Ω'. But the inverse image of this compact set is not compact in Ω, so f is not proper. On the other hand, if f is not proper, we can find a compact K whose inverse image is not compact. Then there is a sequence $\{z_\nu\}$ in $f^{-1}(K)$ that tends to the boundary, while its image stays within K. Thus the condition about sequences fails. Hence this condition is equivalent to f being proper.

$\square$

Suppose $\Omega \subset \mathbf{C}^n$ and $\Omega' \subset \mathbf{C}^N$ are bounded domains. We consider only those holomorphic mappings $f : \Omega \to \Omega'$ that extend continuously to the closure of Ω. That is, we consider only those holomorphic f for which there is a continuous mapping F from the closure of Ω to the closure of Ω', so that f is the restriction of F to Ω.

Suppose $f : \Omega \to \Omega'$ is holomorphic and extends to a continuous mapping of the closures. Proposition VII.4.1 provides a simple way to check whether f is a proper holomorphic mapping. We need only check that f maps the boundary of Ω to the boundary of Ω'. In the special case when the domains are unit balls, this condition becomes $\| f(z) \|^2 = 1$ on $\| z \|^2 = 1$. Our results about squared norms will therefore have applications to proper holomorphic mappings between balls.

Theorem VII.4.2. Let $p : \mathbf{C}^n \to \mathbf{C}^k$ be a holomorphic polynomial mapping, and let $q : \mathbf{C}^n \to \mathbf{C}$ be a holomorphic polynomial. Assume

$$\left\| \frac{p}{q}(z) \right\|^2 < 1$$

for $\| z \| \leq 1$. Then there is an integer N and a polynomial mapping $g : \mathbf{C}^n \to \mathbf{C}^N$ such that $\frac{p \oplus g}{q}$ is a proper holomorphic mapping from $\mathbf{B}_n$ to $\mathbf{B}_{k+N}$. In particular, for $\| z \|^2 = 1$, we have

$$\left\| \frac{p}{q}(z) \right\|^2 + \left\| \frac{g}{q}(z) \right\|^2 = 1. \tag{7}$$

Proof. Consider the polynomial $| q(z) |^2 - \| p(z) \|^2$. By hypothesis it is positive on the sphere. By Theorem VII.3.1 there exists a holomorphic polynomial mapping g such that

$$\| g(z) \|^2 = | q(z) |^2 - \| p(z) \|^2$$

on the unit sphere. Dividing by q yields (7). We may assume, even if p and q were constants, that g is not constant, and therefore $\frac{p \oplus g}{q}$ is not a constant mapping. Its squared norm is not constant but is plurisubharmonic; hence it satisfies the maximum principle. The squared norm equals 1 on the boundary sphere, so it must be less than 1 on the open ball. Therefore $\frac{p \oplus g}{q}$ maps $\mathbf{B}_n$ to $\mathbf{B}_{k+N}$. $\qquad\square$

Corollary VII.4.3. Suppose that q is an arbitrary polynomial that does not vanish on the closed unit ball in $\mathbf{C}^n$. Then there is an integer N and a polynomial mapping p such that at least one component of $\frac{p}{q}$ is reduced to lowest terms and such that $\frac{p}{q}$ maps the unit ball in $\mathbf{C}^n$ properly to the unit ball in $\mathbf{C}^N$.

Proof. Choose any polynomial p_1 such that q and p_1 have no common factors, and such that $| \frac{p_1}{q}(z) | < 1$ on the closed unit ball. For example, cz_1 works for appropriate c. Then, by Theorem VII.4.2, we may find

N and polynomials $p_2, \ldots, p_N$ such that

$$|q(z)|^2 = \sum_{j=1}^{N} |p_j(z)|^2$$

on the unit sphere. The rational mapping $\frac{p}{q}$ satisfies the desired properties.   □

Corollary VII.4.3 relies on Theorem VII.1.1. It is natural to ask whether the result can be proved in a simpler fashion. The author suspects that it cannot, except in the one-dimensional case, where it is elementary! We consider briefly the one-dimensional case. It is easy to show [D] that the proper holomorphic mappings of the unit disk are precisely the finite Blaschke products:

$$B(z) = e^{i\theta} \prod_{j=1}^{k} \frac{a_j - z}{1 - \overline{a}_j z}. \tag{8}$$

In (8) the points a_j lie in the open unit disk, and each a_j can occur with an arbitrary finite multiplicity in the product. Suppose next that q is an arbitrary polynomial in one variable, and that q is never zero on the closed disk. By the fundamental theorem of algebra, q factors into linear factors in $\mathbf{C}$; since its zeroes are outside the unit disk we call them $\frac{1}{\overline{a}_j}$ and see that q is a constant times the denominator in (8). Thus, whenever q is nowhere zero on the closed unit disk, q is the denominator of a proper holomorphic rational mapping of the unit disk. The dimension N from Corollary VII.4.3 can be chosen to equal 1.

Exercise 3. Prove that the proper holomorphic mappings of the unit disk are precisely those mappings satisfying (8). Suggestion: Consider $f^{-1}(0)$. This must be a finite set of points with finite multiplicities. Use it to construct a product $B(z)$ of the form in (8). Then show, using the maximum principle, that $|f| = |B|$.

It is worth noting that, when $n \geq 2$, the (minimal) dimension N needed in Corollary VII.4.3 can be arbitrarily large. In the next exercise the reader is asked to verify this statement for a specific example.

Exercise 4. Let $q(z) = 1 - az_1 z_2$. (Easy) Determine the condition on a so that q is never zero on the closed unit ball. (Fairly difficult) Prove that the minimum integer N that works in Corollary VII.4.3 tends to infinity as $|a|$ approaches the supremum of its allowable values.

VII.5 Positivity on zero sets

What makes a polynomial positive on some set? It is obvious that a quotient of squared norms is always nonnegative. We have been considering versions of a converse assertion. Suppose that a polynomial is positive on some set. Does it agree with a quotient of squared norms there? Our next theorem gives a nice answer to this kind of question.

Suppose that we are given a bihomogeneous polynomial h with zero-set M. Let r be a bihomogeneous polynomial and suppose r is positive on $M - \{0\}$. Then r must agree with a quotient of squared norms on $M - \{0\}$. Theorem VII.1.1 is the special case when we take $h = 0$, and $M - \{0\}$ is the complement of the origin. The proof of the general result is similar to the proof of Theorem VII.2.2.

Theorem VII.5.1. Let M be the zero-set of a (real-valued) bihomogeneous polynomial. Let r be a bihomogeneous polynomial such that $r > 0$ on $M - \{0\}$. Then there are holomorphic polynomial mappings f and g such that

$$r(z, \bar{z}) = \frac{\|f(z)\|^2}{\|g(z)\|^2}$$

for all $z \in M - \{0\}$. We may choose g so that $\|g(z)\|^2$ is a power of $\|z\|^2$.

Proof. By considering the square of the defining polynomial for M, if necessary, we may assume that M is the zero-set of a *nonnegative* bihomogeneous polynomial h. Let S denote the unit sphere. Let η be the minimum value assumed by r on S. Note that η could be negative. Since M is a closed set, and S is compact, $M \cap S$ is compact. Since r is positive there and continuous, its minimum value on $M \cap S$, which we denote by ϵ, is positive. Since r is continuous, there is a $\delta > 0$ such that $r \geq \frac{\epsilon}{2}$ at all points within a distance δ of $M \cap S$.

We claim that there is a $\gamma > 0$ such that, for $z \in S$, $h(z, \overline{z}) \leq \gamma$ implies $r(z, \overline{z}) \geq \frac{\epsilon}{2}$. If this claim were false, then it would fail for each $\frac{1}{n}$. Hence there would be a sequence of points on S such that $h(z_n, \overline{z}_n) \leq \frac{1}{n}$ whereas $r(z_n, \overline{z}_n) \leq \frac{\epsilon}{2}$. Since S is compact, there would be a convergent subsequence satisfying the same inequalities. Call the limit z. By continuity we would have $r(z, \overline{z}) \leq \frac{\epsilon}{2}$, but $h(z, \overline{z}) = 0$. Thus $z \in S \cap M$, and hence $r(z, \overline{z}) \geq \epsilon$. This contradiction proves that such a γ exists.

Now we choose a number C so that

$$\eta + C\gamma \geq \frac{\epsilon}{2}.$$

Given C, we define a bihomogeneous polynomial F_C by

$$F_C(z, \overline{z}) = ||z||^{2a} r(z, \overline{z}) + C||z||^{2b} h(z, \overline{z}).$$

We choose the exponents a and b so that F_C is bihomogeneous. We claim that F_C is positive away from the origin. Since it is bihomogeneous, we need only show that F_C is positive on S. The restriction of F_C to S equals the restriction of $r + Ch$ to S.

Recall that $0 \leq h$. On the sphere where $h \leq \gamma$, we have

$$F_C = r + Ch \geq r \geq \frac{\epsilon}{2}.$$

On the sphere where $h \geq \gamma$, we have

$$F_C = r + Ch \geq \eta + C\gamma \geq \frac{\epsilon}{2}.$$

We have proved that $F_C \geq \frac{\epsilon}{2}$ on S. Since F_C is bihomogeneous, it is positive away from the origin. By Theorem VII.1.1, there is an integer d for which $|| z ||^{2d} F_C(z, \overline{z})$ is a squared norm. Hence

$$|| z ||^{2d} F_C(z, \overline{z}) = || z ||^{2a+2d} r(z, \overline{z}) + C || z ||^{2b+2d} h(z, \overline{z})$$

$$= || A(z) ||^2. \tag{9}$$

Setting $h = 0$ in (9) shows that

$$r(z, \overline{z}) = \frac{|| A(z) ||^2}{|| z ||^{2a+2d}}$$

on the zero-set of h. Since $|| z ||^{2a+2d} = || H_{a+d}(z) ||^2$ is a squared norm, we see that r agrees with a quotient of squared norms on M. This proves the desired result. $\quad\square$

VII.6 Proof of stabilization

To complete the proof of Theorem VII.1.1, we must show that S1) implies S4). To facilitate the proof of this implication, we begin with two lemmas involving integrals arising in the proof and then use a lemma relating positivity and compactness.

Lemma VII.6.1. Suppose that $|| z || < 1$. For fixed positive α, the value of the following function of z depends only upon $|| z ||$:

$$I_\alpha(z) = \int_{\mathbf{B}_n} \frac{|| z - w ||^\alpha}{| 1 - \langle z, w \rangle |^{n+1}} \, dV(w). \tag{10}$$

Proof. It suffices to prove that $I_\alpha(Uz) = I_\alpha(z)$ for each unitary U. We show this using the change of variables formula for multiple integrals. To compute $I_\alpha(Uz)$, make the change of variables $w = U\zeta$ in the integral. The volume form is unchanged because $| \det(U) |^2 = 1$, and the other expressions are independent of U by Proposition IV.1.7. $\quad\square$

Lemma VII.6.2. $\sup_{z \in \mathbf{B}_n} I_\alpha(z) < \infty$.

Proof. By Lemma VII.6.1 we may assume that $z = (z_1, 0, \ldots, 0)$ and $0 \le z_1 < 1$. Let w' denote $(w_2, \ldots, w_n)$. Then

$$I_\alpha(z) = \int_{\mathbf{B}_n} \frac{(|z_1 - w_1|^2 + \|w'\|^2)^{\frac{\alpha}{2}}}{|1 - z_1\overline{w}_1|^{n+1}} \, dV(w).$$

Since $|w_1|^2 + \|w'\|^2 \le 1$, we have

$$I_\alpha(z) \le \int_{\mathbf{B}_n} \frac{(|z_1 - w_1|^2 + 1 - |w_1|^2)^{\frac{\alpha}{2}}}{|1 - z_1\overline{w}_1|^{n+1}} \, dV(w).$$

Now we do the integration in w'. We obtain the volume c_{2n-2} r^{2n-2} of a ball of radius r, where $r = \sqrt{1 - |w_1|^2}$, in $2n - 2$ real dimensions. Thus, for some positive C,

$$I_\alpha(z) \le C \int_{\mathbf{B}_1} \frac{(|z_1 - w_1|^2 + 1 - |w_1|^2)^{\frac{\alpha}{2}}(1 - |w_1|^2)^{n-1}}{|1 - z_1\overline{w}_1|^{n+1}} \, dV(w_1).$$

$$(11)$$

According to Exercise 5 from Chapter I we have the inequalities (12) and (13):

$$1 - |w_1|^2 \le 2|1 - z_1\overline{w}_1| \tag{12}$$

$$|z_1 - w_1|^2 \le |1 - z_1\overline{w}_1|^2. \tag{13}$$

We use (12) and (13) to obtain an integral involving only powers of $|1 - z_1\overline{w}_1|$:

$I_\alpha(z)$

$$\le C \int_{\mathbf{B}_1} \frac{(|1 - z_1\overline{w}_1|^2 + 2|1 - z_1\overline{w}_1|)^{\frac{\alpha}{2}}(2|1 - z_1\overline{w}_1|)^{n-1}}{|1 - z_1\overline{w}_1|^{n+1}} \, dV(w_1)$$

$$\le C' \int_{\mathbf{B}_1} |1 - z_1\overline{w}_1|^{-2+\frac{\alpha}{2}}(|1 - z_1\overline{w}_1| + 2)^{\frac{\alpha}{2}} 2^{n-1} \, dV(w_1)$$

$$\le C'' \int_{\mathbf{B}_1} |1 - z_1\overline{w}_1|^{-2+\frac{\alpha}{2}} \, dV(w_1). \tag{14}$$

To complete the proof we need to show that $I_\alpha(z)$ is bounded for $0 \le z_1 < 1$. Thus it suffices to prove that

$$\sup_{0 \le z_1 < 1} \int_{\mathbf{B}_1} |1 - z_1 \overline{w}_1|^{\eta - 2} \, dV(w_1)$$

is finite for $\eta > 0$.

It is evident that the only possible problem arises as z_1 tends to 1. Therefore it suffices to show that the following integral is finite:

$$\int_{\mathbf{B}_1} |1 - w|^{\eta - 2} \, dV(w). \tag{15}$$

The integrand is nonnegative, and $\mathbf{B}_1 \subset B_2(1)$. The following calculation therefore bounds the integral in (15):

$$\int_{\mathbf{B}_1} |1 - w|^{\eta - 2} \, dV(w) \le \int_{B_2(1)} |1 - w|^{\eta - 2} \, dV(w)$$

$$= \int_{|w-1| \le 2} |1 - w|^{\eta - 2} \, dV(w)$$

$$= \int_{|\zeta| \le 2} |\zeta|^{\eta - 2} \, dV(\zeta)$$

$$= \int_0^{2\pi} \int_0^2 r^{\eta - 1} \, dr \, d\theta = 2\pi \frac{2^\eta}{\eta}.$$

Hence $\sup I_\alpha(z) < \infty$. $\square$

The crucial point in the proof of Lemma VII.6.2 is that the exponent $\eta - 2$ in (15) exceeds -2. The Poisson integral formula, Exercise 2 from Chapter III, shows what happens when η tends to zero.

Remark VII.6.3. One can use the estimate from Exercise 1 from Chapter II to give a similar proof; in both cases one computes the integral over a ball in complex dimension $n - 1$, and ends up needing to show that the same integral is finite for $\eta > 0$.

Lemma VII.6.4. Let T be Hermitian on $L^2(\mathbf{B}_n)$. Suppose that $T = Q + L$, where Q is an operator on $L^2(\mathbf{B}_n)$ whose restriction to $A^2(\mathbf{B}_n)$ is positive definite, and where L is compact on $L^2(\mathbf{B}_n)$. Then there is an integer j_0 such that the restriction of T to $V_j \subset A^2(\mathbf{B}_n)$ is positive definite for $j \geq j_0$.

Proof. We restrict T, Q, and L to $A^2(\mathbf{B}_n)$ without changing notation. Since T is Hermitian, and Q is positive definite (hence Hermitian), their difference L is Hermitian. We have, for some positive number c,

$$\langle Tf, f \rangle = \langle Qf, f \rangle + \langle Lf, f \rangle \geq c\| f \|^2 + \langle Lf, f \rangle.$$

We will show that the conclusion of the Lemma holds when using the positive constant $\frac{c}{2}$ in the definition of positive definite. Suppose otherwise that no such j_0 exists. Then there is an increasing sequence n_j of positive integers and unit vectors $f_j \in V_{n_j}$ such that

$$\langle Tf_j, f_j \rangle \leq \frac{c}{2}\| f_j \|^2.$$

Since the subspaces V_j are orthogonal, the vectors f_j are orthogonal. Combining the previous two inequalities yields

$$\langle -Lf_j, f_j \rangle \geq \frac{c}{2}\| f_j \|^2.$$

This inequality will yield a contradiction. Let $\mathcal{H}$ denote the span of $\{f_j\}$. Because of orthogonality, the inequality implies that $-L$ is positive definite on $\mathcal{H}$. Since $-L$ is compact on $A^2(\mathbf{B}_n)$, it is also compact on $\mathcal{H}$. A compact operator on an infinite-dimensional Hilbert space cannot be positive definite (Exercise 5). This contradiction proves that the required j_0 exists. $\qquad\square$

Exercise 5. Prove that a compact operator on an infinite-dimensional Hilbert space $\mathcal{H}$ cannot be positive definite. (Recall that L is positive definite if there is a positive number c such that $\langle Lf, f \rangle \geq c\| f \|^2$ for all $f \in \mathcal{H}$.) Give an example of a compact operator in infinite dimensions satisfying the weaker statement $\langle Lf, f \rangle > 0$ for all nonzero f in $\mathcal{H}$.

We are now ready to prove Theorem VII.1.1. Recall that the only remaining point is that S1) implies S4). The idea is to use the Bergman kernel as a generating function for the functions $\langle z, \zeta \rangle^d$.

Proof of Theorem VII.1.1. By Theorem III.3.5 the Bergman kernel function for $\mathbf{B}_n$ is given by

$$B(z, \overline{\zeta}) = \frac{n!}{\pi^n} \frac{1}{(1 - \langle z, \zeta \rangle)^{n+1}}.$$

There are thus positive constants c_d such that

$$B(z, \overline{\zeta}) = \sum_{d=0}^{\infty} c_d \langle z, \zeta \rangle^d.$$

Define an integral operator T on $L^2(\mathbf{B}_n)$ by letting its integral kernel be

$$r(z, \overline{\zeta}) B(z, \overline{\zeta}).$$

Observe that T is Hermitian by Remark V.4.3. We will show that $T = Q + L$, where Q and L satisfy the hypotheses of Lemma VII.6.4. Hence the restriction of T to V_j will be positive definite for sufficiently large j. By construction, the restriction of T to V_j is the operator with kernel $c_d \langle z, \zeta \rangle^d r(z, \overline{\zeta})$. Since $c_d > 0$, this yields S4).

Let χ be a smooth function with compact support in the ball, such that $\chi(z) \geq 0$ for all z and $\chi(0) > 0$. We write

$$r(z, \overline{\zeta}) B(z, \overline{\zeta}) = \left(r(z, \overline{\zeta}) - r(z, \overline{z}) \right) B(z, \overline{\zeta})$$

$$+ \left(r(z, \overline{z}) + \chi(\zeta) \right) B(z, \overline{\zeta}) - \chi(\zeta) B(z, \overline{\zeta}). \quad (16)$$

We will consider separately the three terms in (16) to verify that $T = Q + L$ as required.

The explicit formula for $B(z, \overline{\zeta})$ shows that B is singular only where $z = \zeta$ on the boundary. Since χ has support in the interior, their product is smooth and compactly supported. It follows from Theorem V.4.2 that their product is the kernel of a compact operator. Thus the third term in (16) defines a compact operator.

We claim that the second term in (16) is the kernel of an operator whose restriction to $A^2(\mathbf{B}_n)$ is positive definite. Let M_r and M_χ denote the multiplication operators corresponding to these functions, and let P denote the Bergman projection. The operator corresponding to the second term in (16) is therefore

$$Q = M_r P + P M_\chi.$$

For $h \in A^2(\mathbf{B}_n)$ we have $Ph = h$. Also $P = P^*$. Using these facts we compute

$$\langle Qh, h \rangle = \langle (M_r P + P M_\chi)h, h \rangle$$
$$= \langle M_r h, h \rangle + \langle M_\chi h, h \rangle = \langle (M_r + M_\chi)h, h \rangle. \quad (17)$$

Observe that the function $r + \chi$ is positive everywhere on the ball. Hence the last expression in (17) is larger than $c|| h ||_2^2$. Thus Q is positive definite on $A^2(\mathbf{B}_n)$.

It remains to study the first term in (16); to finish the proof it suffices to show that this term is the kernel of a compact operator. We derive compactness from the following result, where $\overline{\mathbf{B}}_n$ denotes the closed unit ball in $\mathbf{C}^n$ and B is the Bergman kernel function. Note that $L^2(\mathbf{B}_n)$ and $L^2(\overline{\mathbf{B}}_n)$ are the same Hilbert space.

Lemma VII.6.5. Let $A : \overline{\mathbf{B}}_n \times \overline{\mathbf{B}}_n \to \mathbf{C}$ be a continuous function such that there are positive constants C and β such that

$$| A(z, \zeta) | \leq C |z - \zeta|^\beta \quad (18)$$

for all $z, \zeta \in \overline{\mathbf{B}}_n$. Put $K(z, \zeta) = A(z, \zeta)B(z, \overline{\zeta})$. Then K is the integral kernel of a compact operator S on $L^2(\mathbf{B}_n)$.

Proof. We use Proposition V.2.3. Given $\epsilon > 0$, we will show that there is a compact operator S_1 so that the following estimate holds:

$$|| Sf || \leq \epsilon || f || + || S_1 f ||.$$

Recall from Lemma VII.6.2 that, for each $\alpha > 0$, the supremum of $I_\alpha(z)$ over $z \in \mathbf{B}_n$ is finite. We denote this supremum by $m(\alpha)$.

Consider $\delta > 0$, choose $z \in \mathbf{B}_n$, and let Ω_δ denote the intersection of $\mathbf{B}_n$ with the set where $|z - w| \leq \delta$. Let χ denote a smooth function of ζ, supported in Ω_δ, such that $0 \leq \chi \leq 1$ and $\chi = 1$ in some smaller neighborhood of z.

For a fixed z we write S as $S = \chi S + (1 - \chi)S = S_0 + S_1$. The kernel $(1 - \chi)K$ of S_1 is continuous on $\overline{\mathbf{B}}_n \times \overline{\mathbf{B}}_n$, because $1 - \chi$ vanishes near the set where $z = \zeta$ and the Bergman kernel is smooth away from this set. By part 4 of Theorem V.4.2, S_1 is compact.

Suppose we can show that $\| S_0 \| \leq \epsilon$. Then

$$\| Sf \| = \| S_0 f + S_1 f \| \leq \| S_0 f \| + \| S_1 f \| \leq \epsilon \| f \| + \| S_1 f \|,$$

and S will be compact by Proposition V.2.3.

To verify that $\| S_0 \| \leq \epsilon$, we use Theorem V.4.1. Let K_0 denote the kernel of S_0. We will find $\delta > 0$ so that, for all z:

$$\int_{\mathbf{B}_n} | K_0(z, \zeta) | \, dV(\zeta) \leq \epsilon.$$

The corresponding inequality for all ζ follows in the same way.

Using the properties of χ and $A(z, \zeta)$ we have

$$\int_{\mathbf{B}_n} | K_0(z, \zeta) | \, dV(\zeta) = \int_{\mathbf{B}_n} | \chi(\zeta) K(z, \zeta) | \, dV(\zeta)$$

$$= \int_{\mathbf{B}_n} | \chi(\zeta) | \, | A(z, \zeta) | \, | B(z, \overline{\zeta}) | \, dV(\zeta)$$

$$\leq \int_{\Omega_\delta} | A(z, \zeta) | \, | B(z, \overline{\zeta}) | \, dV(\zeta)$$

$$\leq \int_{\Omega_\delta} C | z - \zeta |^\beta | B(z, \overline{\zeta}) | \, dV(\zeta)$$

$$\leq C \delta^{\frac{\beta}{2}} \int_{\Omega_\delta} | z - \zeta |^{\frac{\beta}{2}} | B(z, \overline{\zeta}) | \, dV(\zeta)$$

$$\leq C \delta^{\frac{\beta}{2}} m \left(\frac{\beta}{2} \right). \tag{19}$$

We therefore choose δ such that $C\delta^{\frac{\beta}{2}}m(\frac{\beta}{2}) \leq \epsilon$ and the string of inequalities in (19) yields the required estimate. Therefore S is compact. □

To complete the proof of Theorem VII.1.1 we need to show that the first term in (16) defines a compact operator. That term is $(r(z, \overline{\zeta}) - r(z, \overline{z}))\, B(z, \overline{\zeta})$; we claim that the hypotheses of Lemma VII.6.5 hold with $A(z, \zeta) = r(z, \overline{\zeta}) - r(z, \overline{z})$. This claim is easy to see; $A(z, \zeta)$ is a polynomial in the variables $z, \overline{z}, \overline{\zeta}$ and vanishes when $\overline{\zeta} = \overline{z}$. We can therefore write

$$A(z, \zeta) = \sum_{j=1}^{n} a_j(z, \overline{z}, \overline{\zeta})(\overline{z}_j - \overline{\zeta}_j)$$

where the a_j are also polynomials, and hence continuous. Then (18) holds with $\beta = 1$. Therefore Lemma VII.6.5 applies, and the first term in (16) is the kernel of a compact operator. We have finally proved all the needed statements for Theorem VII.1.1. □

We close the book by providing some insight into the proof that the first term in (16) is the kernel of a compact operator. The idea is simple: the only singularities in the Bergman kernel function arise when $z = \zeta$; the numerator $r(z, \overline{\zeta}) - r(z, \overline{z})$ vanishes when $z = \zeta$ and thus compensates for the singularity there.

This argument has a nice formulation using commutators. Recall that $[A, B]$ denotes the commutator $AB - BA$ of operators A and B. The operator coming from the first term in (16) can be written

$$\sum c_{\alpha\beta} M_{z^\alpha}[P, M_{\overline{z}^\beta}]. \tag{20}$$

In the commutator $[P, M_{\overline{z}^\beta}] = [P, M]$, we multiply by $\overline{\zeta}^\beta$ in the term PM, whereas we multiply by $\overline{z}^\beta$ in the term MP. The reason is that ζ is the variable of integration in the formula for P. The sum in (20) is finite, and each M_{z^α} defines a bounded operator on $L^2(\mathbf{B}_n)$. By Corollary V.2.7, (20) will be compact if each operator R_β, defined by

$$R_\beta = [P, M_{\overline{z}^\beta}] = PM_{\overline{\zeta}^\beta} - M_{\overline{z}^\beta}P,$$

is compact. The kernel of R_β is

$$\frac{n!}{\pi^n} \frac{\overline{\zeta}^\beta - \overline{z}^\beta}{(1 - \langle z, \zeta \rangle)^{n+1}},$$

and the argument used in the proof of Lemma VII.6.5 shows that R_β is compact.

The same idea applies much more generally. See [CD2] for a result on the compactness of $[P, M]$ where P is the Bergman projection of a domain much more general than the unit ball, and M is a more general operator.

Remark VII.6.6. The intuitive idea is that the commutator of bounded integral operators with reasonably nice kernels should be *better* than bounded. This principle holds rather generally and should continue to find applications throughout analysis.

CHAPTER VIII
Afterword

This afterword aims to provide some glimpses into research about positivity conditions in complex analysis. The papers mentioned here form an incomplete list, but perhaps will give the interested reader an entry to the research literature.

This proof of Theorem VII.1.1 gives a nice application of the theory of compact operators to a concrete problem about polynomials. The proof here follows the approach from [CD1], where Theorem VII.3.1 also appeared. Theorems VII.4.2 and VII.5.1 come from [D2]. The main assertion in Theorem VII.1.1, that S1) implies S3), goes back to [Q]; the proof there uses the Hilbert space $A^2(\mathbf{C}^n, dG)$ mentioned in Example II.1.6. There is no explicit mention of compact operators in [Q], although some of the key estimates are closely related to compactness. As here, one of the key points is to estimate a commutator. The main application in [Q] is a proof of the projective Nullstellensatz, which implies the usual Nullstellensatz. Quillen points out that his direct analytical proof of the Nullstellensatz resembles the proof of the fundamental theorem of algebra using Liouville's theorem.

It is possible to reformulate Theorem VII.1.1 as a theorem about Hermitian metrics on powers of the *universal line bundle* over *complex projective space*. See [CD3] and [D3]. This reformulation then generalizes to an isometric imbedding theorem ([CD3]) for holomorphic vector bundles over compact complex manifolds. This imbedding theorem evokes Calabi's result [C] on isometric imbedding, as well as

243

Kodaira's imbedding theorem. See [W] for a detailed treatment of the Kodaira imbedding theorem and also general information about complex manifolds and holomorphic vector bundles.

The proof of the general isometric imbedding theorem in [CD3] is similar in spirit to the proof of Theorem VII.1.1 given here. The Bergman kernel, commutators, compact operators, the version of the Cauchy-Schwarz inequality from Chapter VI, and other ideas from this book all appear in the proof of the general result. See also [CD2] for a generalization of Lemma VII.6.5. The expository paper [D3] considers many of the topics in Chapters VI and VII; the positivity conditions from Definition VI.5.1 come from [D5] but also appear in [D3]. The material on logarithmic plurisubharmonicity comes from [D4]. We do not attempt here to give references on the vast subject of pluripotential theory. See [Fe] for results relating bordered complex Hessians and the Bergman kernel function.

Proposition VI.1.3 and Pólya's Theorem VI.1.4 are special cases of Theorem VII.1.1, so one can also express them in terms of line bundles. In Proposition VI.1.3, for example, the given polynomial p can be as a metric on a power of the *universal line bundle* over complex projective space $\mathbf{CP}_1$. The function $x \mapsto 1 + x$ gets reinterpreted as the Euclidean metric on the first power of the universal line bundle. Raising it to a high power amounts to taking a high tensor power. See [D2] for more discussion of this example.

Theorem VII.3.1 is one of our main applications of Theorem VII.1.1. There are several distinct but related results along the following lines. Suppose a function r is positive on the unit sphere, or more generally, on the boundary M of a nice domain in $\mathbf{C}^n$. Must r agree on M with the squared norm of a holomorphic mapping f? See [L1], [L2], and [L] for positive results of this type. These important papers solve this problem in situations differing from the setting of this book. Here we have assumed that r is a polynomial and we want f also to be a polynomial.

This book is about *complex analysis*. The reader might be wondering about corresponding results in *real analysis*. In his address to the International Congress of Mathematics in 1900, Hilbert posed twenty-three problems. Here we mention only the seventeenth problem: Is a nonnegative polynomial in n real variables necessarily a sum of squares of rational functions? Exercise 7 of Chapter 6 provides a simple example of a nonnegative polynomial that cannot be written as a sum of squares of *polynomials*.

Artin solved Hilbert's 17th Problem in 1926; he proved that the answer is yes, although his proof is not constructive. His techniques led to significant developments in *real algebra* and in logic. The recent book [PD] provides a current and complete treatment of positivity conditions for real polynomials; it begins with the definition of an ordered field. Artin's proof and many subsequent developments appear in detail. The authors make useful bibliographical and historical comments at the end of each chapter.

We close by offering a few additional thoughts on positivity conditions. Definition IV.1.2 gives the meaning of *nonnegative definite* for a bounded operator L on a Hilbert space. Such an operator always has a square root; that is, $L = A^*A$ for some A. Suppose $L(z, \bar{z})$ is a nonnegative definite operator for each $z \in \mathbf{C}^n$. For each fixed z we can find a square root. Under what circumstances can we take a square root that varies holomorphically with z? Finding such a family of square roots is not always possible even when L is a scalar and $L(z, \bar{z})$ is a polynomial in z and $\bar{z}$. The isometric imbedding result from [CD3] provides a situation where we can multiply an operator by a nice scalar function such that the resulting operator does have a holomorphic square root. Finding nice square roots of operator-valued functions is often important in applied mathematics, engineering, and physics. See [RR]. The references [B] and [Dj] also provide interesting information on closely related topics.

This terse appendix provides some basic information on the prerequisites for this book. It consists of definitions and statements of some concepts and results used, but not developed, in the book.

A.1 Algebra

Definition. (Group) A *group* is a mathematical system consisting of a set G, a binary operation, $(g, h) \mapsto gh$, and a distinguished element 1, such that the following hold:

1) (1 is the identity element.) $1g = g1 = g$ for all $g \in G$.

2) (Existence of inverses). For all $g \in G$, there is a (necessarily unique) $g^{-1} \in G$ such that $g^{-1}g = gg^{-1} = 1$.

3) (Associative law). For all $f, g, h \in G$, we have $(fg)h = f(gh)$.

Mathematicians generally say things such as "Let G be a group." This is an abbreviation for something more precise such as "Let G be a set, with a binary operation on G and an element 1 in G satisfying the three axioms in the Definition." This type of abbreviated language applies to fields, vector spaces, and other mathematical systems. The brevity gained more than compensates for the precision lost. For example, with this language the letter G stands for both the group and the

247

underlying set of elements. One also says "G is a group under multiplication"; this clause has a clear meaning, yet it blurs the issue whether G means the set or the mathematical system.

A group is called Abelian (or commutative) if $gh = hg$ for all g and h in G. Often one writes $g + h$ instead of gh when the group is Abelian. This applies especially in cases where there is another operation considered as multiplication.

Definition. (Field) A *field* is a mathematical system consisting of a set F, two binary operations called addition and multiplication, and two distinguished elements 0 and 1, such that axioms 1), 2), and 3) are true. We write addition as $(x, y) \mapsto x + y$ and multiplication as $(x, y) \mapsto xy$.

1) F is an Abelian group under addition with identity element 0.

2) $F - \{0\}$ is an Abelian group under multiplication with identity element 1.

3) For all $x, y, z \in F$ the distributive law holds:

$$(x + y)z = xz + yz.$$

We write the additive inverse of x as $-x$, and the multiplicative inverse of a nonzero x as $\frac{1}{x}$ or x^{-1}.

Definition. (Ordered field) An *ordered field* is a mathematical system consisting of a field F and a subset P of F with the following properties. We call P the set of positive elements in F.

1) 0 is not in P.

2) Suppose $x \neq 0$; then either x is in P or $-x$ is in P.

3) 1 is in P.

4) If $x, y \in P$, then $x + y$ and xy are in P.

Statement 2) does not allow the possibility that both x and $-x$ are in P. In particular, if $x = -x$ for some x in a field, then the field cannot be ordered.

There are examples of fields that admit more than one ordering; therefore the term *ordered field* includes the choice of the subset P.

Let F be an ordered field. We write $x > y$ if and only if $x - y \in P$. We write $x \geq y$ if and only if either $x = y$ or $x > y$. We write $x < y$ if and only if $y > x$, and we write $x \leq y$ if and only if $y \geq x$. The axioms for an ordered field imply the usual rules for manipulating inequalities.

There are many examples of fields. The rational number system $\mathbf{Q}$ and the real number system $\mathbf{R}$ are ordered fields; in each case the positive set consists of positive numbers in the usual sense. Finite fields cannot be ordered. We illustrate this in a special case. Let p be a prime number, and let $\mathbf{Z}_p$ denote the set $\{0, 1, \ldots, p-1\}$ of ordinary integers. We make $\mathbf{Z}_p$ into a field by performing arithmetic and multiplication modulo p. The numbers 0 and 1 are the identity elements. The additive inverse of 1 is $p - 1$. If the field were ordered, $p - 1$ would be positive as it is the sum of this many copies of 1. On the other hand, it would be negative as the additive inverse of the positive element 1. The assumption that an ordering exists thus leads to a contradiction.

Definition. (Vector space) A *vector space* V over a field F is a mathematical system consisting of an Abelian group V and a function $F \times V \to V$, written $(c, v) \mapsto c \cdot v$ and called scalar multiplication, such that axioms 1) through 5) hold. Let $\mathbf{0}$ denote the additive identity in V and let 0 denote the additive identity in F. Let 1 denote the multiplicative identity in F. Then:

1) $0 \cdot v = \mathbf{0}$ for all $v \in V$.

2) $1 \cdot v = v$ for all $v \in V$.

3) $c \cdot (v + w) = c \cdot v + c \cdot w$ for all $v, w \in V$ and all $c \in F$.

4) $(c_1 + c_2) \cdot v = c_1 \cdot v + c_2 \cdot v$ for all $c_1, c_2 \in F$ and all $v \in V$.

5) $c_1 \cdot (c_2 \cdot v) = (c_1 c_2) \cdot v$ for all $c_1, c_2 \in F$ and all $v \in V$.

Elements of V are called vectors, and elements of F are called scalars. Generally (and in this book) one denotes 0 and **0** by the same symbol.

A vector space is called *finite-dimensional* if there is a finite subset $\{v_1, \ldots, v_n\}$ of V such that the following holds. For each $v \in V$ there are elements $c_j \in F$ such that

$$v = \sum_{j=1}^{n} c_j v_j.$$

Such a sum is called a *linear combination* of the v_j.

A.2   Analysis

It is a bit artificial to distinguish between algebra and analysis. For example, one might say that *fields* belong to algebra while *ordered fields* belong to analysis. Certainly inequalities are a major feature of analysis. It is the completeness axiom for the real numbers that gets analysis going.

We recall the standard definition of completeness for the real number system. Let F be an ordered field, and let S be a nonempty subset of F. Then S is bounded above if there is some $m \in F$ such that $x \leq m$ for all $x \in S$. Such an m is an *upper bound* for S.

Definition. An ordered field F is complete if every nonempty subset $S \subset F$ that is bounded above has a least upper bound in F.

The least upper bound of S is also called the *supremum* of S; it need not be in S. The supremum of S is the unique m_0 such that $m_0 \leq m$ for every upper bound m of S.

Up to isomorphism, the real number system **R** is the unique complete ordered field. The rational number system **Q** is an ordered field, but is not complete. The complex number system **C** is a complete field, but cannot be ordered.

A normed linear space is a vector space where it makes sense to talk about the *length* of a vector. More precisely we have the following definition:

Definition. (Normed linear space) Let V be a vector space over F, where either $F = \mathbf{R}$ or $F = \mathbf{C}$. We say that V is a *normed linear space* if there is a real-valued function $|| \ ||$ on V, called the *norm*, such that:

1) $||v|| > 0$ for all nonzero v in V.

2) $||cv|| = |c| \ ||v||$ for all $c \in F$ and for all $v \in V$.

3) $||v + w|| \leq ||v|| + ||w||$ for all v and w in V.

As a consequence of 2), we have $||\mathbf{0}|| = 0$.

The simplest examples are of course $\mathbf{R}^n$ and $\mathbf{C}^n$ with the Euclidean norms as defined in the text. It is easy to prove, but never needed in this book, that all norms on a finite-dimensional vector space are equivalent, in the sense that they lead to the same topology. This means that they define the same collection of open sets. It suffices for us to recall the notion of open set in the metric space setting. Metric spaces have *distance functions* whose properties model those of normed linear spaces. When V is a normed linear space, we define a function $d : V \times V \to \mathbf{R}$ by $d(u, v) = ||u - v||$. This function measures the distance between two points in V. More generally we have the following definition.

Definition. (Metric space) A *metric space* (M, d) is a mathematical system consisting of a set M and a function $d : M \times M \to \mathbf{R}$ (called the distance function) such that

1) $d(x, y) > 0$ if $x \neq y$.

2) $d(x, x) = 0$ for all $x \in M$.

3) $d(x, y) = d(y, x)$ for all x and y in M.

4) $d(x, z) \leq d(x, y) + d(y, z)$ for all x, y, z in M (the triangle inequality).

Often one writes M instead of (M, d), especially when the distance function is clear from context. Normed linear spaces provide good examples of metric spaces.

Definitions. The open ball $B_r(x)$ of radius r about x is the subset of M consisting of all y for which $d(x, y) < r$. The closed ball $\overline{B}_r(x)$ of radius r about x is the subset of M consisting of all y for which $d(x, y) \leq r$.

Definitions. A subset $A \subset M$ is called open if, for each $x \in A$, there is a positive number ϵ (generally depending on x) so that $B_\epsilon(x) \subset A$. A subset $K \subset M$ is called closed if its complement is open.

It is elementary to verify that an open ball is open and a closed ball is closed.

Definitions. A sequence $\{x_v\}$ in a metric space (M, d) converges to x (or "has limit x") if, for each $\epsilon > 0$, there is a positive integer N so that $n \geq N$ implies $d(x_n, x) < \epsilon$.

A sequence $\{x_v\}$ in a metric space (M, d) is a *Cauchy sequence* if, for each $\epsilon > 0$, there is a positive integer N so that $n, m \geq N$ implies $d(x_n, x_m) < \epsilon$.

Convergent sequences are always Cauchy sequences; the converse assertion holds in *complete* metric spaces.

Definition. A metric space (M, d) is *complete* if every Cauchy sequence in M has a limit in M.

Definition. A *Banach space* is a complete normed linear space.

The next two lemmas make the connection between closed sets and completeness.

Lemma. A subset K of a metric space (M, d) is closed if the following is true: whenever $\{x_\nu\}$ is a sequence in K whose limit x exists in M, then necessarily $x \in K$.

The next lemma holds in particular for closed subspaces of Banach and Hilbert spaces, and is used several times in the book.

Lemma. A closed subset K of a complete metric space (M, d) is itself a complete metric space (with the same distance function).

Many results from elementary real analysis fit into the general framework of metric spaces, and the usual epsilon-delta arguments from elementary calculus generally work in this setting. Limit and convergence arguments acquire a certain elegance. Early in the book there is some detail when these items arise, but especially later some of this standard material is used without reference. The reader might consult [Ah], [F], and [TBB] for more information.

We continue with a list of terms the reader should know, and then close the appendix with a brief discussion of some convergence results.

Definitions the reader should know in the metric space setting: compactness, completeness, connectedness, continuous function, dense set, equicontinuous family of functions, sequence, subsequence, uniform convergence, uniformly continuous function.

Let f be a bounded real-valued function on a set A; the set of values $f(A)$ is then a non-empty subset of **R** that is bounded above. Therefore it has a supremum m. Several times in this book, and very often in pure and applied mathematics, one needs to know that f achieves its supremum somewhere. Thus we seek an x with $f(x) = m$. The next result gives us circumstances when this is possible.

Theorem. (**Min-Max Theorem**) Suppose that (M, d) is a compact metric space, and $f : M \to \mathbf{R}$ is a continuous real-valued function. Then f is uniformly continuous, f is bounded, and there are points in M at which f achieves its infimum and its supremum.

The next result makes use of a metric space of functions. Let (M, d) be a compact metric space, and let $C(M)$ denote the linear space of continuous complex-valued functions on M, with norm given by

$$\|f\| = \sup_{x \in M} |f(x)|.$$

Then $C(M)$ is a Banach space. We are interested in describing its compact subsets. To do so, we briefly discuss *equicontinuity*.

Let $\mathcal{F}$ be a family of continuous functions on a metric space (M, d). For each $f \in \mathcal{F}$, and for $x \in M$, the usual epsilon-delta definition provides the meaning for the phrase "f is continuous at x." In general there is no reason to expect, for a given ϵ, to be able to choose δ independently of x or of f. In case (M, d) is compact, each $f \in \mathcal{F}$ is uniformly continuous, so we can then choose δ independently of x for a fixed f. Equicontinuity is the analogous idea, where we want to choose δ independently of f.

Definitions. A collection $\mathcal{F}$ of continuous complex-valued functions on a metric space (M, d) is *equicontinuous at x* if, for each $\epsilon > 0$, there is a $\delta > 0$ so that

$$d(x, y) < \delta \Rightarrow |f(x) - f(y)| < \epsilon$$

for all $f \in \mathcal{F}$. We say that $\mathcal{F}$ is *equicontinuous* if it is equicontinuous at each $x \in M$. Finally $\mathcal{F}$ is *uniformly bounded* if there is a constant C such that $|f(x)| \leq C$ for all $x \in M$ and for all $f \in \mathcal{F}$.

Theorem A.2.1. (**Arzelà-Ascoli**) Suppose that M is a compact metric space and $\mathcal{F}$ is a family of continuous complex-valued functions on M.

Then $\mathcal{F}$ is a compact subset of the Banach space $C(M)$ if and only if the following three conditions hold:

1) $\mathcal{F}$ is equicontinuous.
2) $\mathcal{F}$ is uniformly bounded.
3) $\mathcal{F}$ is closed.

Bibliography

[Ah] Ahlfors, Lars V., *Complex Analysis*, McGraw-Hill, New York, 1979.

[An] Andrews, Larry C., *Special Functions of Mathematics for Engineers*, McGraw-Hill, New York, 1992.

[B] Bellman, Richard, *Introduction to Matrix Analysis*, McGraw-Hill, New York, 1970.

[BFS] Boas, Harold P., Fu, Siqi, and Straube, Emil, The Bergman kernel function: explicit formulas and zeroes, *Proc. Amer. Math. Soc.*, Vol. 127 (1999), 805–811.

[C] Calabi, Eugenio, Isometric imbedding of complex manifolds, *Annals of Math* 58 (1953), 1–23.

[CD1] Catlin, David W. and D'Angelo, John P., A stabilization theorem for Hermitian forms and applications to holomorphic mappings, *Math Research Letters* 3, (1996) 149–166.

[CD2] Catlin, David W. and D'Angelo, John P., Positivity conditions for bi-homogeneous polynomials, *Math Research Letters* 4, (1997) 1–13.

[CD3] Catlin, David W. and D'Angelo, John P., An isometric imbedding theorem for holomorphic bundles, *Math Research Letters* 6, (1999) 1–18.

[D] D'Angelo, John P., *Several Complex Variables and the Geometry of Real Hypersurfaces*, CRC Press, Boca Raton, 1993.

[D1] D'Angelo, John P., An explicit computation of the Bergman kernel function, *Journal of Geometric Analysis*, Vol. 1 (1994), 23–34.

[D2] D'Angelo, John P., Positivity conditions and squared norms of holomorphic polynomial mappings, *Contemporary Mathematics*, Volume 251 (2000), 163–172.

[D3] D'Angelo, John P., Proper holomorphic mappings, positivity conditions, and isometric imbedding, to appear in Proceedings of KSCV4.

[D4] D'Angelo, John P., Bordered complex Hessians, *Journal of Geometric Analysis*, Vol. 11, Number 4 (2001), 559–569.

[D5] D'Angelo, John P., Positivity conditions for real-analytic functions, Complex Analysis and Geometry, *Ohio State Univ. Math Res. Inst. Publ* 9, de Gruyter, 2001.

[Dj] Djokovic, D. Z., Hermitian matrices over polynomial rings, *J. Algebra* 43 (1976), 359–374.

[Do] Donoghue, William F., Jr., *Distributions and Fourier Transforms*, Academic Press, New York, 1968.

[Fe] Fefferman, Charles L., Monge-Ampere equations, the Bergman kernel, and geometry of pseudoconvex domains, *Annals of Mathematics*, 103 (1976), 395–416.

[F] Folland, Gerald B., *Real Analysis*, John Wiley and Sons, Inc., 1984.

[FK] Folland, G. B. and Kohn, J. J., The Neumann Problem for the Cauchy-Riemann Complex, *Annals of Mathematics Studies* 75, Princeton University Press, Princeton, 1972.

[GK] Greene, Robert E. and Krantz, Steven G., *Function Theory of One Complex Variable*, Wiley Interscience, New York, 1997.

[HLP] Hardy, G. H., Littlewood, J. E., and Polya, G., *Inequalities*, Cambridge Univ. Press, London, 1934.

[H] Hörmander, Lars, *An Introduction to Complex Analysis in Several Variables*, North Holland Publishing Company, Amsterdam, 1973.

[Hu] Huylebrouck, Dirk, Similarities in Irrationality Proofs for π, $\ln 2$, $\zeta(2)$, and $\zeta(3)$. *American Math Monthly*, Volume 108 (2001), 223–231.

[Ka] Katznelson, Yitzhak, *An Introduction to Harmonic Analysis*, Dover Publications, New York, 1976.

[Kr] Krantz, Steven G., *Function Theory of Several Complex Variables*, Wadsworth and Brooks/Cole, Belmont, California, Second Edition, 1992.

[L1] Lempert, László, Imbedding Cauchy-Riemann manifolds into a sphere, *International Journal of Math* 1 (1990), 91–108.

[L2] Lempert, László, Imbedding pseudoconvex domains into a ball, *American Journal of Math* 104 (1982), 901–904.

[L] Løw, Eric, Embeddings and proper holomorphic maps of strictly pseudoconvex domains into polydiscs and balls, *Math Z.* 190 (1985), 401–410.

[PD] Prestel, Alexander and Delzell, Charles N., *Positive Polynomials: From Hilbert's 17th Problem to Real Algebra*, Springer-Verlag, Berlin, 2001.

[Q] Quillen, Daniel G., On the representation of Hermitian forms as sums of squares, *Inventiones math.* 5 (1968), 237–242.

[R] Reznick, Bruce, Some concrete aspects of Hilbert's 17th Problem, *Contemporary Mathematics*, Volume 253 (2000), 251–272.

[RR] Rosenblum, M. and Rovnyak, J., The factorization problem for nonnegative operator valued functions, *Bulletin A.M.S.* 77 (1971), 287–318.

[Ru1] Rudin, Walter, *Function Theory in the Unit Ball of* $\mathbf{C}^n$, Springer-Verlag, New York, 1980.

[Ru2] Rudin, Walter, *Functional Analysis*, McGraw-Hill, New York, 1973.

[SR] Saint Raymond, Xavier, *Elementary Introduction to the Theory of Pseudodifferential Operators*, CRC Press, Boca Raton, 1991.

[S] Strichartz, Robert S., *A Guide to Distribution Theory and Fourier Transforms*, CRC Press, Boca Raton, 1994.

[Ta] Taylor, Michael E., *Partial Differential Equations: Basic Theory*, Springer, New York, 1996.

[TBB] Thompson, Brian S., Bruckner, Judith, and Bruckner, Andrew, *Elementary Real Analysis*, Prentice-Hall, Upper Saddle River, 2001.

[W] Wells, Raymond O., *Differential Analysis on Complex Manifolds*, Prentice-Hall, Englewood Cliffs, New Jersey, 1973.

Index

261